The Year of the Red Deer

By the same author

HIGHLAND YEAR
HIGHLAND DEER FOREST
WILD HIGHLANDS

Lea MacNally

The Year of the Red Deer

with 100 monochrome photographs and 8 pages of colour plates
by the author

London: J. M. Dent & Sons Ltd

First published 1975

Made in Great Britain
by Biddles Ltd.
Guildford, Surrey
for
J. M. DENT & SONS LTD
Aldine House, Albemarle Street, London

ISBN 0 460 04157 6

Dedication

For my boys, Lea and Michael, who aided and enriched so much of my deerwatching by their company and interest

Contents

Illustrations

Colour Plates

Foreword

This does not pretend to be a scientific work on red deer, simply a recording of a lifetime's watching and photographing Highland red deer. I hope it may lead to a better understanding of these, the most splendid wild animals we have left in Britain today. Red deer *(Cervus elaphus)* are the largest of our British deer and the main wild population lives in the Highlands of Scotland. Other areas where there are smaller pockets of red deer are Galloway in Scotland, and the Lake District, Exmoor, Devon and Somerset, Thetford Chase in Norfolk and a few in the New Forest. This does not include those in various 'parks' in England. All these last-mentioned deer are usually larger of body and antler than our Highland deer, purely due to environment, as also are the red deer of the European continent. One may regard this difference as simply a result of geographical situation and modification in the one basic species. The North American counterpart of our red deer is the wapiti *(Cervus canadensis)*.

That the red deer *are* still with us in the Highlands is a tribute to the fact that they have adapted, over the centuries, to become the species best fitted to the spartan existence of life on the high and often bleak mountains where they spend most of their year. No other form of land use utilizes the *high* ground in the Highlands as well and as *cheaply* as do red deer. The pity is that in so many areas of the Highlands today the lower sheltered glens which the deer use mainly when weather conditions force them to, in winter and early spring, which are then essential to their well-being, are being progressively denied to them by the commercial planting of close-ranked conifers. In the Highlands forestry can use land only up to about 1,500 feet if an acceptable crop is to be obtained, and the practice requires state aid. Red deer use the land up to the very mountain tops. Their management is *not* state aided, while their value as meat-producing animals increases yearly, quite apart from their attraction and enhancement of hills which, without them, would be sterile indeed.

Those other users of hill land in the Highlands, hill sheep and cattle, also receive state aid. It is undeniable that, like forestry, they must play a part in land use in the Highlands, *but so also must red deer.* It is gradually being realized that more far-sighted planning is required if we are to strike a happy medium of co-existence between these four methods of land use in the Highlands. But meanwhile still more of our lower glens are going under a rash of drab, dark green conifers. Some day, perhaps, we may learn.

Finally, I must record an immense debt to the kindness and help of too many stalkers and proprietors to mention individually. The continued existence of deer owes a lot to them also, and will continue to do so; for there is no one with a higher regard for, indeed, at times, utter dedication to, 'his' deer, than the true Highland stalker.

Hind with her calf in June

A three-week-old calf asleep

A stag in the red coat of summer, his antlers in velvet

A hind at her 'looking glass'

At rest in the August heather

Browsing on alder leaves

September: for the calf (below), its first autumn

Two fine stags, one at ease, the other alert

Stag with hinds at the rutting time

A hind in winter coat, in December snow

Stags in the winter herd

A sixteen-pointer, a very fine Highland stag, in March, shortly before the antler-casting time

Summer RENAISSANCE TO THE ROARING-TIME

June in the Highlands, a vital irresistible month of fresh scents, sounds and scenes, a month truly alive with the activity of new life and new growth, species vying with species in the endless struggle that is Nature. A well-worn cliché you may feel, but, well worn as it may be, absolutely true.

As one walks up a wooded river glen, en route to the hill, a shrill, long-drawn-out 'mewing' cry may draw one's gaze upwards to watch, against the clear blue of the sky, a buzzard, hunting to feed a nestful of hungry young. Two or three leisurely wingflaps, a soaring wheel, a long glide on wide-set wings, again two or three flaps, a wheel then another long glide, keen eyes intent on the landscape spread below. A pair of mistle-thrushes 'churr-churr' in rattling alarm as the intruder passes, too near for their liking, their well-camouflaged lichened nest crammed to bursting point with four nearly-fledged young; a goosander 'quarck-quarcks' harshly as it speeds down river toward its nest in a hollow alder tree, almost shaving the water as it flies, following every twist and turn of the current; a·tiny willow warbler flutters out at one's feet and perches anxiously yet confidingly a scant yard away while one experiences anew the fascinated admiration always aroused by the daintiness and seeming fragility of its neat domed nest and almost transparently-white eggs, minutely dotted with red. Rowan and bird-cherry flourish, white-bedecked, amid the lovely fresh green of the birch leaves and the duller green of hazel; grass is at long last replacing its winter yellow with a welcome green while, less welcome but thriving regardless, the head of new bracken uncoils itself, reptile-like, green from a mattress of decaying rusty-red. On some favoured slopes bluebells are beginning to tinge the green with steely-blue and, even daintier and more fragile-seeming than the willow warbler's nest, a cluster of wood-sorrel, exquisite, delicate, white violet-veined flowers embowered in a sheltering cranny cleaving the decayed bark of a fallen birch tree, adds its beauty, for those with eyes to see, to the June scene.

Weaving among the birches, outstretched in desperate flight, no time for airy, graceful bounds, a young roebuck, rusty-red in new coat, suddenly appears, intent only on keeping out of reach of the older, savagely-barking buck that is equally intent on ridding 'his' territory of the presumptuous youngster he is pursuing; neither deigns to notice the enthralled onlooker.

Less obvious, but spotted by the quick eye of the silent walker, a huge wildcat, boldly black-ringed, blunt-ended tail extended stiffly behind her, so engrossed in her own affairs as to be temporarily unalert, stalks stealthily above the river, braving the hostile daylight in quest of prey to meet the ever-hungry demands of her litter of spitting furies, secure in their rock-slit den in a crevice of a sheer river cliff.

Farther out, into the hill ground now, main river and trees left behind; overhead a couple of narrow-winged young ravens already strong in flight cronk-cronk in unison while, higher still, cleaving the air on scimitar-like wings, three swifts hawk for invisible insects. Gliding with infrequent flap of wing, equally at ease whether against or with the wind and so high he appears hardly bigger than the swifts, a golden eagle hunts in seeming leisurely flight that will alter dramatically to an air-splitting swoop whenever prey is spied. His mate, far off on their bulky eyrie, is rending apart, with audible 'scrunch' of bone, a cock grouse obscenely nude of feather except for red-combed head, from which she proffers, with incongruous gentleness, bloody morsels to her young, as yet still clad entirely in white down though already weighing nearly four pounds. A hen grouse (perhaps 'widowed' by the eagle for she is unsupported by a cock in her maternal anxiety) suddenly explodes, 'clucking' in her trepidation, out of the apparently bare heather at one's feet and a moment later a sparrow-sized ball of yellow-brown fluff rises hazardously out of the same heather, beating the air valiantly with little paddles of wings, to pitch into the heather again about thirty yards distant while the mother, almost demented now, snakes and weaves

In early June single hinds may be seen away from the herds, an indication that their calving time is near.

underfoot in frantic distraction display, a dauntless exhibition of selfless maternal courage, needlessly displayed did she but know it. In a ridge-top gully well above the heather line the betraying taint of human, carried ahead on the light breeze, sends four sandy-yellow fox cubs scurrying from their macabre play with the skull of a dead lamb into the cover of a cairn of huge grey rocks, while the vixen, lying amid a huge apron of scattered scree, seemingly dispassionate but securely overlooking her den, watches the unsuspecting human pass below.

All this, and much more besides, is June and it is against this backdrop of bustling parental activity and juvenile wants that life comes anew to the red deer herds, quietly and without fuss, on ground so remote from human interference, at this

Fatalities in giving birth, as in this picture, are extremely rare with red deer.

time of year, that relatively few people in Britain have ever seen a new-born red deer calf or even know when or where they are being born. The current (1973) estimate of the number of red deer in Scotland is 185,000, of which by far the majority live in the Highlands. There is a relatively small number in Galloway, small in number but high in quality for they live in a better habitat than that of most Highland red deer.

My own experience is mainly of Highland red deer and here few calves are born before June. There are exceptions of course, an occasional calf being born before the end of May and, at the other extreme, as late as early October. These very late calves are rare, and since the normal gestation period for red deer hinds is eight months the mothers of such calves must have come into season and mated about February instead of the more normal October. This can be accounted for by the fact that a red deer hind normally comes into season for eighteen to twenty-four hours, usually in October, and if she is not mated with then (though this is rare), will go off season and come on again eighteen to twenty-one days later for a similar period, repeating this, as I know from personal

A newly born red deer calf, its coat still damp and dark from its birth, is not a beauty. . . .

experience, until as late as February, or until she *is* mated with.

The actual birth of her calf rarely causes a red deer hind much trouble, far less than is the case with the ewes of the hill sheep for instance, where lambing shepherds may have to go round their flocks twice daily at lambing-time to succour those in difficulties. On an agricultural programme broadcast regularly on the Scottish Home Service, called 'Farm Journal', on 21st March 1973 mortality figures were quoted that surprised me, of 5 per cent of lambs stillborn, 20 per cent dying in the first few days of life and a further small percentage dying before a fortnight old. In marked contrast to this I have found only three hinds dead while trying to give birth, in twenty-five years of experience of hill red deer, and in five of these years I was actively engaged in searching for red deer calves on the hill for the purpose of personal research. In those five years I found only four dead calves, three of which had died 'naturally' and one had been killed by an eagle. Over this same period 131 calves were found, caught, ear-tagged, weighed and sex noted, 70 of the 131 being female and 61 male. The average weight of recently-born calves was found to be 14-15 lb. It intrigued me, yearly, to find that the sex ratio

A few hours later it is transformed into a charmingly dappled creature of boundless attraction.

A hind uttering the 'calving bellow', a sound heard only at the calving season.

almost invariably levelled out even when it seemed to be trending to predominance in one or other sex. For instance in one year I had four successive hind calf captures, but the next four captures were all stag calves. The following year the reverse occurred when I had five successive stag calf captures, and the next five were all of hind calves. I found a similar near equality in the annual calf mortality, which takes place, a ruthless pruning-out of the weaklings, about mid March to mid April, when the calves are nine to ten months old.

A phenomenon of the calving season in red deer which I have never found mentioned in the extensive literature on them is the sound made by the hinds *only* at that season. This I eventually came to call 'the calving bellow'. I first heard this noise, rather like the juvenile bellow made by an immature stag in emulation of his elders and betters, while en route at 5 a.m. one Sunday morning to the eyrie of a golden eagle. It mystified me completely but the mystery was solved on the following Sunday while I was going at the same early hour to the same eyrie. At this second time of hearing I was lucky enough to see as well as hear the noise-maker: a hind, head and neck outstretched and mouth open as if baying the moon while she emitted a doleful bellow, pregnant with a palpable air of unrest and yearning—fanciful as this may seem to some. That was in 1961, and every year since then in the calving season I have subconsciously listened for and invariably heard the 'calving-bellow' on at least a few occasions. In my initial inexperience I formed the opinion that it was made by a hind under stress in the imminence of the birth of her calf, but this was only partly correct. By the time I had finished my self-imposed five years' research on deer calves (which called for my daily presence on the hill throughout the daylight hours of the twenty-one days which constitute the main calving time), I had heard and seen the 'calving-bellow' used so often and variously that I had to revise my thinking. I have now known it used before the birth of a calf, during the actual birth, after the birth (in this case once with an air of what I can only describe as smug self-satisfaction by a hind lying on the bare black of a wide peat hag, beside her newly born calf), in mourning after the death of a calf (in three of the four recently-born calf deaths I found, I saw and heard the mother bellowing near the dead calf), and once it was being employed by a hind whose calf was trapped deep in an underground peat runner.

This calf, apparently lured by that peculiar attraction which peat seems to have for a variety of reasons for all ages of Highland deer, had gone in at the open drain-like end and worked its way forward, ever deeper and farther in, and eventually was literally in a tunnel of peat which was so circumscribed that when I had to negotiate it to rescue the calf, with a very definite, cringing unwillingness of both body and spirit, I had to worm through its Stygian darkness on my stomach, stripped down just to shirt and trousers. At the farther end of this tunnel in the peat there was a small underground chamber in which the calf crouched, held as by a magnet by the glimmer of light from a small hole which pierced the surface some four feet above. The hind, either calling above or periodically thrusting her nose in at the hole, undoubtedly helped to keep the calf's attention on the impossible 'exit' above instead of letting her find, even if by accident, the way she had gone in. A distinct well-worn track to the edge of the hole testified to the hind's admirable if mistaken devotion. It *may* be that the calf would have found her own way out eventually; on the other hand it is quite likely that the lucky combination of my presence on that part of the hill and the bellowing of the mother, which drew my attention from about a mile away, saved her life. My one regret is that I had no companion that day, not only as insurance in case I got stuck, but to photograph the filthy, black, peat-encrusted stalker who reversed, belly-flat, out of the peat tunnel dragging by the hind legs an equally filthy, loudly protesting and lustily struggling calf not in the least grateful for its rescue from the tomb. It was a hind calf of 26½lb. and probably about ten days old.

A pre-conceived notion that deer calves are born mostly during the night was also shattered with experience. I found that calves might be born at any hour of the day and probably night, and on areas of the hill ranging from glen bottoms to, at the other extreme, the high-ground daytime lying-up places. This is the reason why a hind seen apparently alone in an area of the hill normally devoid of deer in the June daytime, under normal weather conditions, is indicative of imminence of birth or of the presence of a recently born calf lying hidden some-

Foxes will occasionally kill a red deer calf if they find one when the mother is absent.

The golden eagle is also a predator on young deer but usually, like the fox, only in the mother's absence.

Wildcats have sometimes been accused of killing young red deer calves but this must be very exceptional.

where near her. *Unless she is disturbed by human approach* a hind will remain near her recently born calf in its first few hours of life, though in doing so she may be tormented (as I have personally seen, and shared) almost beyond her endurance by midges and other winged pests from which she could have escaped by going to the breeze-cooled tops with her 'sisters'. A hind so stationed, at this very vulnerable period of her calf's life, will defend it against fox or eagle, the two main predators on young calves, so resolutely and selflessly as to rout them almost without exception. Only from mankind will she run, a rather revealing reflection on humanity, groundless though her fears are in most cases. This is not to say that a certain toll is not taken each year by fox and eagle who themselves have hungry young to feed in June. I have seen, often enough, remains of deer calf at fox den or eagle's eyrie. Pathetic, and both pity- and anger-evoking, as are these dappled remains of a youngster with such abundant appeal, which had barely known life, it must be realized that in nature predators, *of whom man is the chief,* are as entitled to existence as herbivores. Let us remember too that predators kill mainly to feed themselves and their young; not for profit, hatred or because it is politic. In some cases, undoubtedly, the remains seen at den or eyrie are of calves found dead, for neither fox nor eagle will scorn such 'easy meat'. In other cases, just as undoubtedly, the calf has been killed, usually because the mother has grazed a bit too far from the young calf. Unpalatable as it is to think thus I also believe that at times this is the result of disturbance by humans; not necessarily an approach near to a calf but simply human presence, often unavoidable, sufficient to clear an area of hill of any hinds keeping vigil near young calves. Neither eagle nor fox needs long to make its kill, and even though the attacked calf may have time to emit that ear-splitting, high-pitched panic-stricken squeal which is guaranteed to bring *every* mother hind within earshot galloping from a hill apparently empty a moment before, it is usually too late. The outraged mother will often rout either eagle or fox from the body of her calf and will stand over or near it for perhaps many hours after, but eventually she will drift away, memory in nature, in cases like this, being mercifully short.

As to the actual birth of a calf, I have been lucky enough to witness this more than once now, as I have described in an earlier book. The first birth I ever saw, however, is indelibly impressed on my memory, and I recorded this event at the time:

'The air of the June morning was sultry and the deer had been moving out since dawn, from their lower-ground feeding to the cooler high ground. By 8 a.m. they were lying in scattered groups on the coolness of damp peat hags or on the exposed ridges, cudding contemplatively, or just dozing.

'For one hind, however, heavy in calf, the coolness of the peat on which she lay afforded little solace as she grew increasingly restless, feeling the stirring of new life inside her. Twice already since she had arrived in the high coire, its face still patched with stubborn drifts of hard-packed snow, she had shifted position, leaving each time, beside the impress of her body on the soft peat, a moist, white-streaked exudation, evidence to the experienced eye that her "time" was not far off.

'A large, dark hind, by 9 a.m. she was too restless to lie any longer and, leaving the company of the three hinds with whom she'd been lying she began to walk stiffly around the semicircular face of the green coire, tail held rigidly erect as if straining to part company with her body. Emerging unseen from the cover of a gully, I immediately sat down to spy at her. She had a distraught air about her and the peculiar manner in which she held her tail confirmed my suspicion that she was near to calving.

'As I lay watching, opposite her and almost on her level, but about a mile away, she traversed the coire, at times pausing to kick out spasmodically with one or other hind leg, at others nuzzling quickly, as in protest, at her haunch. Her progress was interrupted periodically as she lay down, and each time I thought she would calve, but each time she rose after a few minutes and resumed her painful progress around the coire's arc. When she had left her erstwhile companions a good half mile behind she lay down for a longish period and it seemed certain that there she would calf but, after an obviously uncomfortable rest of some fifteen minutes, up she got again, this time, being now near the bounds of the coire, to retrace her steps. Her mode of progression was now altered, at times she made a quick run forward, as if suddenly stung into action, but once again she punctuated her "walks" with frequent rest periods.

'The belief I formerly held that a hind always goes away on her own to calve was now exploded as the hind rejoined her former companions and lay down heavily, as if with a sigh of relief, just below them. It seemed definite now that her time was near; she elevated head and neck and gave vent to that yearning, mournful, almost bovine bellow which the normally silent red deer hinds give vent to at the calving season. Lying full length now, on her side, she rubbed head and neck hard along the ground, as if wallowing. At 10.30 a.m. she jumped quickly to her feet and I saw, as she stood broadside, a full six inches of two black-looking legs, with white-edged hooves, protruding from her. Again she lay down, on her opposite side, and went through the whole performance of leg twitchings, head nuzzling, bellowing and lying at full length, rubbing head and neck on the ground. To all this the hinds nearby paid no heed whatever, continuing placidly to chew the cud.

'The calving hind lay longer this second time but when she arose again I could see that she was a little further advanced. A third time she lay down, changing sides again, this time immediately at full stretch, head and neck vigorously scrubbing at the grass. An obviously strong heave, and simultaneously she literally sprang to her feet and the calf, glistening perceptibly, dropped from her; I winced involuntarily as it hit the grass. The time was 11.17 a.m.; she had been calving for over two hours, and had traversed a mile of hill in the process.

'Immediately, the mother began to tear at the glistening envelope which enclosed the calf, her head jerking as she tore at it strongly, and as she raised her head a long streamer of white hung from her jaws. Some of this she swallowed, some discarded. The black-looking damp head of the calf could be seen, wobbling uncontrollably at first, but gaining strength visibly as the hind licked at it assiduously. As she did so one of the nearby hinds, curious now, rose and walked across, head and neck held low, stiffly extended well in front of her, as if to sniff at the strange new arrival. The mother declined to co-operate in this and without halting her cleaning of her calf she adroitly contrived to shoulder the nosy hind away. She, on her part, did not press her quest for "knowledge" but turned away to graze, leaving the hind to continue her solicitous licking.

'By 11.25 the calf was striving to rise, without much success, but by 11.37 it was actually on its legs, only to collapse immediately. It persevered in its struggles to rise, however, and in its desire to get below the hind to suck. The hind was amazingly clumsy, to my eyes,

The recently born calf, in its efforts to find the hind's udder, more than once tried at the 'wrong end'.

and kept swinging round, time after time knocking her calf sprawling in its pathetic blunderings near to her. Once, in fact, when the calf succeeded in making its first temporary contact with her udder she shied away as if stung and this time really sent the youngster flying. In these first purely instinctive efforts to find the source of sustenance, as often as not the calf was seeking between the forelegs instead of the hind legs of the mother. It was 11.45 when it found the udder again and, now in fair control of its legs, began avidly to draw in the life-giving milk, only to collapse almost at once as if the rich warm milk had anaesthetized it. Moments later it was up again and now unerringly made contact; this time it sucked strongly for six minutes, after which, its small stock of strength apparently for the moment exhausted, it folded up again.

'Satisfied that the calf had now had its necessary first feed, I began to walk across to ear-tag and weigh the calf. The nearby hinds ran first; the mother, understandably, waited until I was very near before her fears overcame her maternal devotion, and then followed suit, while the calf didn't even seem to notice her departure. The calf was a female and, at an hour old, was already almost dry with the dapples strongly marked on her coat. Lighter than average in weight at that age, she weighed only 12lb., and at that early stage of her life she had no fear of humans, which is quite typical of very young deer calves. As I sat for a moment or two enjoying the sight of the appealing little creature she rose on legs still wobbly and staggered over to a nearby peat bank, where she began to mouth at the crumbly dry peat. This was of intense interest, for only a few days before I had done a post-mortem on a dead day-old deer calf and had found a hard-packed "golf-ball" of peat inside it. I left her, moments later, lying again now, in the shelter of the peat bank.

'Out of sight again, watching from my former viewpoint, I noted with satisfaction the return of the hind, soon enough to make me surmise she had been watching all the time, herself unseen, from a

A life-giving drink at last, while the hind simultaneously licks her calf.

A hind 'greeting' her calf in the evening after a day-long absence on the higher ground while her calf has stayed behind.

A red deer calf soon learns to sample grass and heather to supplement its mother's milk.

By the time it is ten days old the calf will accompany its mother wherever she goes, and is fleet enough of foot to outdistance any human.

distance. I watched, with rather wry amusement, her initial recoil as she sniffed at her calf and detected the obviously repugnant scent of man. Maternal instinct, however, over-rode instinctive fears. Another searching sniff, despite an involuntary shrinking, disclosed to her sensitive nose that this was indeed her calf and she allowed its eagerly questing mouth to find her udder.

'Satisfied, I slid back, and resumed my rounds, speculating when or where, if ever, I would next come across the calf I had seen come into the world, while the calf, untroubled by any such speculation, continued greedily to suck. The mother, her restlessness and labour pangs alike assuaged, let the calf have its fill and later left it lying, while she wandered off some distance to graze, eventually going out of sight. It would be evening before she returned and by that time the calf would be strong enough to come a little to meet her, the events of the morning already forgotten.'

Twins are so rare with Highland red deer as to have been unrecorded, with *definite* proof, until two research workers on the Island of Rhum, Fiona Guinness and John Fletcher, were lucky enough actually to witness the birth of twins to a red deer hind (at the 'wrong' time of the year incidentally) on 9th October 1970. Admittedly she was a hand-reared hind but she was living under near natural conditions in a very large enclosure of heather and peat with a number of other hand-reared hinds and the stag who had sired her twins. Both calves weighed 13 lb. 12 oz. at birth. Personally, I had always been convinced that twins could and did occur with Highland hinds

Twin red deer are almost unknown in the Highlands but I have on occasion seen two calves with a hind in circumstances which led me to believe they were twins.

though extremely rarely. On three occasions in twenty-five years I had seen two calves with a hind in circumstances that made me sure they were twins. Few people of course would accept this as probable, the oft-repeated theory in a case like this being that the second calf had been an adopted, orphaned one. Those who theorize thus might think again if they had seen as many calves as I have mistakenly approach a hind which was not their mother and be invariably and most savagely repulsed by a flailing fore-hoof, so that at times I actually feared for their survival (in one instance I saw a calf hit so hard on its fragile skull that it was stunned momentarily). Though it is by no means always wise to draw parallels between domestic stock, even hill sheep, and wild animals, any shepherd will tell you how difficult it is to persuade a ewe which has lost its lamb to take an orphaned one in its place. The usual but not always successful procedure is to skin the dead lamb and loosely tie the skin around the live orphan so that the distinctive scent of the ewe's *own* dead young one masks, or at least mingles with, the *different*, to her, scent of the orphan, until such time as the body scent of the foster mother becomes transferred to the fosterling. This also applies to hill cattle. In view of all this I remain convinced that twins are *very occasionally* born to wild Highland red deer.

Another point of disagreement I have with 'popular' beliefs about red deer concerns the again oft-quoted one that a hind, when alarmed with a very young calf in attendance, nudges it

Red deer calves that made history, the twins whose birth was actually witnessed on the island of Rhum.

with her nose, touches it with an admonitory fore-hoof, or otherwise signals it to drop and lie doggo while she runs and leaves it. In all the many times I have seen hinds run thus, *never* was any kind of signal whatever given to the calf. In some cases the calf stood momentarily bewildered, watching the departing hind, before dropping of its own volition and lying still; in others the calf wobbled unsteadily a short distance before collapsing to lie still, and in the case of slightly older calves, say twelve to twenty-four hours old, they ran some distance behind the mother, dropping at last as their weakening energies left them far behind the fleeing hind. As calves get older they consciously seek cover before dropping, in a peat hag, amid long heather, by a sheltering rock, or in the runnel of a shallow hill burn; but a very young calf drops anywhere, sometimes where there is no cover whatever. It might be a bare green, or in very short heather, or even on the yet more conspicuous bareness of a black peat flat. In such cases the reputed camouflage of the deer calf lying still is non-existent and even to human eyes they are glaringly obvious. How much more so to fox or eagle?

By July the strain of months of carrying their unborn calves is over for most of the calf-bearing hinds, and for most of the stags the strain on their constitution of growing their new antlers is also nearly over. For both sexes, in fact, July might be

Camouflage gone wrong: born in the bareness of the black peat hag, this red deer calf, a few hours old, is only too easy for fox or eagle to see.

The hind has all the parental cares; a hind (right) with her calf of the year and of the previous year—hind, calf and follower in Highland parlance.

In June, while the hinds are giving birth to their calves, the new antlers of the stags are growing rapidly beneath their covering of velvet.

likened to a month of relative peace and plenty, vicissitudes of winter and spring forgotten. The mother hinds of course still have the parental 'worries' of their youngsters, while the stags, in whose make-up, polygamous as they are, parental cares do not figure at all, still have their antlers, growth completed as it may be, enveloped in velvet.

The typical Highland July sight of red deer is of a herd mainly of one sex or the other (for the sexes, in the case of mature red deer, spend most of the year in separate herds, except for the mating time) lying on the ridges of the high ground, coats now gleaming red and immaculate, contrasting beautifully with the green of the ridge. On a day of sun and white fleecy cloud, in particular, such a sight can be absolutely breathtaking. Some will be lying placidly chewing the cud, some dozing, only the occasional twitch of an ear indicating life. Here and there a beast will be lying sound asleep curled up like a dog, nose to rump with not even the twitch of an ear. Only when deer feel *absolutely* secure will they sleep thus; one sees them much more often relaxed and dozing but still incredibly watchful, in various attitudes besides the normal 'chewing the cud' recumbent position with all four legs tidily tucked in beneath them. They may be seen lying stretched out on their sides, all four legs extended restfully; or with three legs tucked under and one foreleg extended; less frequently with legs tucked under them and head and long neck outstretched, 'chin' resting on the ground, eyes closed, but not *sound* asleep as in the dog-like curled-up attitude.

In marked contrast to the indolence of their elders will be the antics of the livelier calves still attractively clad in dappled

By late June antler growth is well advanced and most of the stags will be in their coats of summer red.

In early July the antlers of many of the mature stags are fully developed, but some of the young staggies (as on the right) may show very little antler growth.

coats. While their lazier contemporaries lie near their mothers (many more of them trustingly in the securely asleep curled-up attitude than in the case of their vigilant elders), these livelier dappled sprites may be gambolling about like spring lambs though with infinitely more grace and appeal, playing 'king of the castle', one of them perched on a knoll butting at the others seeking to depose it; or one moment grazing desultorily, already learning the differing tastes of the hill vegetation, the next, racing madly away, zig-zagging here and there, leaping rocks, drains and other obstacles, to 'turn on a sixpence' and race as madly back, resuming grazing as abruptly as it had ceased. Yet another, suddenly infected by irresistible *joie de vivre,* may begin a series of incredibly fast bucking bounds, curving, sunfishing, swapping ends, coming down on all four stiffly extended rigid legs, adept as any bronco of wild west fiction. Such entrancing intimate glimpses are unforgettable, the more so for being rarely witnessed, out on the high grounds as they are, for the greater part of the human race lacks opportunity, or, given the opportunity, the energy and skill to penetrate to the summer sanctums of Highland red deer. This is probably the most carefree period of red deer life, this calf stage, with no knowledge of the unsettling presence of 'the master stag' keeping each hind group in almost ceaseless turmoil for the period of the rut or of the hardships of the winter ahead, just enjoyment of life of the moment, secure in the vigilance of their elders.

Red deer stags on a high peaty plateau, antlers still in velvet though now fully grown, summering apart from the hind herds.

A surprising aspect of the hind herds until well into July is that new calves form a much smaller proportion of the number than one expects. On one mid July day I counted two herds of deer, one of 108 which had only 26 calves with it, the second of 107 with only 22. The reason, I believe, is that even by mid July there are still calves too young, or in some cases too lazy, to follow their mothers when they leave their nightly grazing areas down in the glens for the daytime peace of the high ground. This is probably a purely Highland phenomenon, for with Highland deer there is a very marked and often lengthy daily movement; in the evenings down to the better grazing and shelter of the glens, in the mornings back out again to their high-ground solitudes where most often their only observer in July is the golden eagle.

Deer do not embark on lengthy daily peregrinations needlessly (unlike some humans) but neither do they like risk of disturbance by humans or the hot weather we sometimes get even in the Highlands. Red deer in the Highlands logically

In the sultry days of July any deer that remain about the birch woodlands of the lower glens may seek coolness and shade in the cover of the bracken under the trees.

Many deer take to lying on the ridges to get the cooling breeze and to escape the attentions of the various flies that are a torment on the lower ground.

enough are much better adapted to endure cold than heat, much better than is the human observer. The moist, humid mild weather we more often get, when every blade of grass and sprig of heather seems to spawn midges, clegs, warble-flies and nasal bot-flies, drives the deer almost frantic (and the first two insect pests named similarly afflict human beings also on occasion). Ears twitch incessantly, eyes roll as the deer endeavour to keep track of their tiny persecutors, heads go down to be 'pawed' at by the hoof of a foreleg, until they are goaded beyond endurance or are stung, often literally, into

A young hind, sleek in her summer coat, alarmed by the unwelcome attention of parasitic flies and ready to rise and take speedy evasive action.

Red deer become accustomed to water from an early age and are excellent swimmers, neither rivers nor lochs being a barrier to them.

mad evasive flight from their tormentors, flight that is seldom successful. A wild, twisting race ensues, preceded perhaps by a series of bucks lifting hindquarters high in the air and often ending in the attacked deer scrubbing its nose frantically in the wiry heather. It is for these reasons that deer desert the more fruitful but heat-trapping and fly-producing grazing of the glens for the quiet and relative coolness of the tops, where the breezes, however light, help to dispel heat and flies alike. In autumn, winter and spring with little or no heat, no flies, little grazing on the tops and breezes no longer gentle, deer are much less inclined to desert the lower areas, quite apart from the threat of inclement weather which will bring them off the tops irrespective of the time of year.

In many parts of the Highlands where there are huge areas of high ground, peaty plateaus cut into mazy formation by zig-zagging peat hags, meandering black-looking burns and the occasional inky-black water of a peat-surrounded lochan, the deer find a different kind of sanctum from heat and flies. In hot or humid weather I have watched hinds and calves, at first grazing slowly toward me, vanish in the growing heat of the day one by one into drain-like peat hags, the heather-grown banks concealing all but top of head and twitching ears, even these disappearing when their owners sought the ultimate in

On very hot days coolness may be sought by lying in the pools of the hill burns . . . a shake or two after rising dispels most of the water from a deer's coat.

coolness by stretching head and neck along the peaty bottom of the hags. A heather flat which one could have sworn could *not* hide even one deer conceals completely a group of twenty or so, the emptiness being magically transformed, as one encroaches, into fleeing deer as the hags spew forth their occupants. Their elegant, richly-red coats of summer are stained and daubed, dripping with moist black peat, but the deer are unconcerned; indeed, one might think that they

Stags as well as hinds avail themselves of the solace of the cooling waters of lochan or burn in the heat of late summer.

derive as much enjoyment from this as any mud-pie-making urchin, and perhaps with more reason, for there is no threat of cervine parental retribution. At other times I have watched deer subside, as with a sigh of relief, into the placid dark water of a pool in a burn, only head, neck and line of back visible, this last awash in the gently flowing current; then rise when completely satiated, and saturated, and expel, with one vigorous shake, the soothing wetness in an enshrouding nimbus of silvery droplets. On occasion the peaty lochans are used to achieve coolness, a scattering of deer fringing the shallows, belly deep, heads drooping as if to get as near to the coolness radiating from the surface as possible, now and again dipping muzzles in to drink briefly; for deer *do* drink water quite frequently, despite the theory I have seen advanced that they get sufficient moisture in their grazing to preclude the necessity for this.

To the stags, with antlers still in velvet, the flies may be even more of a torment than to the hinds. I have seen a July stag, antlers velvet-clad though full grown, and still bulbous in appearance at the tops, with these same bulbous tops so thickly clustered with flies that one could have sworn there was no room for even one more. From notes taken in mid July of another year, I quote:

A stag with the velvet stripping from his antlers can look very untidy. Rubbing the antlers vigorously on trees, bushes or even through coarse tufts of rushes or heather soon cleans off most of the velvet and darkens the antlers in the process.

'A group of a dozen stags, antlers in velvet, were lying, strangely enough, on a ridge in the low ground of the hill. The group included a ten-pointer and an eight-pointer which I recognized from the previous year, their heads a facsimile of last year's. Four others were mature stags, the remaining six, younger ones. Most incongruous in this all male gathering was a small dappled calf which appeared to have adopted a young staggie as 'foster mother'. No hinds were within miles of the area that day. The stags were having very little peace from the flies brought out by the sultriness of the day, which made it the more peculiar that they had not made for the high ground. They were never at rest, constantly shifting position, throwing up their heads with a quick jerking heave every so often, or lying with heads stretched out straight before them, nostrils as close as they could get to the cool heather and even cooler dampness of the mosses. Within a period of half an hour I saw three of them obviously attacked by nasal bot-flies. One was very badly affected, the first sign being a violent blowing down his nostrils, clearly audible to me as I lay concealed about a hundred yards distant. He rose at last and ran a few yards shaking his head wildly and frantically blowing through his nostrils in very obvious pain, now and

By mid August the antlers of many of the mature stags will be completely free of velvet.

The newly exposed antlers are white-looking tinged with pink.

again rubbing his nose into the heather or dabbing at it with a fore-hoof. After some minutes he stopped this apparently unavailing evasive action and stood still, twisting his head this way and that, nose pointing skyward, and still blowing through his nostrils in his bid to eject the tiny tormentors in his tender nasal passages. Finally he stampeded, bucking and prancing, till he was out of sight behind the ridge and it was a good ten minutes before he reappeared, now apparently more at ease.'

Though the relative quiet of the summer hills begins to be shattered in many areas by the opening of the grouse season on 12th August, and later by the start of the stalking of the stags on some of the 'earlier' deer forests, the pattern of behaviour is much the same throughout August and for most of September. The adult deer are probably looking at their very best at this period, immaculate in the thin red coat of summer, the red coat from which they get their name, shorter-lived though it is than the thicker darker coat of late autumn and winter which displaces it. It is usually well into July before the majority of the deer are in summer red and by early October most of them will be back in dark winter dress.

By mid August many of the mature stags have their new antlers clean of velvet, at first white looking, tinged pink with blood left by the shedding of the velvet, darkening later as the stag rubs them on trees, posts, clumps of rushes or heather and uses them to aid his trampling fore-hooves in stirring up a peat wallow. A stag not completely clean, tatters of velvet hanging from his points, looks comical, a cervine scarecrow, but he seldom remains long like that. I suspect that the velvet shed off the antlers is often eaten by the stag himself, probably for the same reason that Highland deer 'chew' antlers, bones of defunct sheep and deer and even alien objects like pieces of alloy fabric of crashed aeroplanes—a craving for mineral traces largely deficient in the acidity of the peat-based grazing which forms the major part of their habitat. The velvet seems to take about three weeks to a month to dry off and 'die' after the antlers' full growth is completed. Its dying off is believed to be caused by the bone of the coronet (the 'corrugated' rim or 'collar around the base of the antler where it emerges from the pedicles on the head of the stags) growing outwards imperceptibly but inexorably as the growth of the antlers nears completion, and consequently stretching thinner and thinner the enclosing velvet until it at last is worn through by the jagged edges of the coronet. The velvet, cut through all around at its base and deprived of its nourishment, like a ring-barked tree, dies, and when it is completely dead and dry seems to become a source of irritation to the stag so that he is moved to rub his hitherto jealously guarded, tender, growing antlers vigorously on whatever 'rubbing' material is at hand. Lacking trees on much of his summer high-ground habitat, this may be willow scrub, tufts of coarse rushes, a centuries-old bleached stump imbedded in an expanse of peat, relic of the Caledonian forest, or even the posts or wire of a high-ground march fence. If the last mentioned is in a state of disrepair, as many of the ancient high-ground fences inevitably are, relics of the past in fact, the stag may end up with a crown of rusty wire, or worse still, if the wire retains sufficient strength, may be snared and held by it until starvation ends his imprisonment. The most harrowing instance I know of wire entanglement was of a youthful stag that was found dead one March with a coil of stranded wire so tight around head and antlers that it had grooved into one antler, into the bony rim of one eye socket and deep into the lower jawbone on the same side. Having eased its tension by grooving vertically deep into the lower jawbone it had remained static at the end of its groove, but through the cud-chewing motion, daily necessary for the stag's survival, it had worn a circular hole about the size of a 5p. (shilling) piece at the end of the groove. The torture that stag must have endured in clinging to existence defies imagination or description, lasting as I believe it did from late August or early September until the following March. He still had a dried hard wisp of velvet on one top and I believe he had been using the wire of an old march fence (miles away from where he was found dead) to aid in cleaning the velvet from his antlers and had been snared by it. How long it took him to break free must remain a matter for speculation but how hard he had to struggle may be gauged by the grooving of the wire into antler and skull.

By early August the calves are strong and vigorous, capable of following their mothers over the roughest terrain.

By late August or early September, depending on their birth date, most of the calves of the year will have lost much of their appeal with the disappearance of their dappled coat and the gradual growing in of their first winter coat. At the distance from which wild red deer will usually be looked at, with

When the calves are about three months old the dapples begin to fade from their coats and soon disappear completely.

telescope or binoculars, the calves' sex will be almost indistinguishable for, under Highland habitat conditions, a young male will be over a year old before the two hair-covered knobs on his head, where in later years the antlers will grow, are readily distinguishable. The term indeed for a Highland male red deer of this age is the self-explanatory one of a 'knobber', though I have heard people who should have known better apply it to a young stag with his first 'spikes' of around nine inches in length, as the average first *antlers* will be in the Highlands. Under optimum conditions (e.g. in deer parks or in good woodland habitat) young males may grow these spikes while the Highland male is only producing knobs, a year ahead in terms of antler growth. Those who are critical of Highland red deer because of this retarded antler growth and, usually, lesser weight and size, should remember however that Highland deer are an example of natural selection in successful adaptation to a poor environment of bare hill from that of sheltered woodlands; natural selection which took place over centuries and meant that the hardiest and best fitted to cope with rigorous conditions survived, losing weight and size inevitably in their long period of adaptation, but

In September the stags are at the peak of condition, fat after the summer months of good grazing.

By late September a growing restlessness is apparent in the stag herds.

A stag travelling in search of hinds is absolutely obsessed by his desires; any other stag met now is a potential rival.

retaining a grace and hardihood second to none of the cervine race in Britain or Europe.

Throughout most of September the stags, still in their herds, continue to live in relative amity, but by the third week of the month the restlessness and increasing ferment of the approaching rutting time is beginning insidiously to infect them. Roaring, that compelling, evocative, spine-tingling voice of the red deer rut, will begin to echo amid the high tops while most of the stags are still together; then, day by day, stag after stag will depart, without ceremony or farewell until, with the coming of October, practically all the mature stags are engaged in seeking hinds. When, then, a stag meets a fellow stag with which he has perhaps grazed or lain side by side only days before, it will be as a potential rival for female company, and the first rule of the red deer rut, 'might is right', will decide which keeps or loses the hinds.

Stag after stag will leave the herds in which they have spent the summer, to seek out the hinds.

The traditional 'day of the roaring' (its starting date) in the Highlands is 26th September.

Autumn THE ROARING-TIME

In the Highlands, to the Gael the traditional 'day of the roaring' is 26th September and undoubtedly in most years roaring, on the forests where the stag herds are, has started by then, even though the rut, as the mating-time of the red deer is known, seldom gets into full swing before 1st October. Stags and hinds spend most of their year in separate herds, perhaps miles apart, and the deer forest that is large enough and of suitable terrain to hold both is fortunate indeed. On a stag-forest, roaring and progressive unrest mark the onset of the rut, while day by day the herds dwindle as the stags go their separate ways in seeking hinds. Tantalizingly for those who own a hind-forest, the onset of the rut will be heralded by the distant roaring, from well over the march, of stags as yet more vocal than active, until, on one unforgettable day, the first stag will be seen with the higher-ground hinds. From then on every day will see more and more stags in, while roaring, especially towards dusk and dawning, reaches such a crescendo that the very hill seems to echo and re-echo with it. Competition for the hinds waxes fierce, hinds who have no say whatever in choosing their 'lord and master', a lord and master, moreover, who may change a few times during the course of the rut, for it takes a real 'master' stag, in terms of strength and weight, to keep his hinds for the three weeks in October that mark the height of the rut.

Eyes wildly rolling, a stag roars his challenge to the hills at the rutting time.

The first stag to 'come in' may be *with*, rather than *hold* (roaring continuously and possessively as he will be) an enormous herd of hinds and calves. I remember in one year counting a herd of 179 with the first stag to come in, but as more and more stags come in and competition for hinds increases so do the big herds inevitably get split up into more manageable groups. A rutting stag, holding hinds, will not

At the height of the rut the quiet of the hills is shattered by challenge and counter-challenge.

tolerate *any* other male in his herd older than the male calves of that year. The males of the previous years, knobbers to around sixteen months old, are summarily evicted from the family group, bewildered, one can only surmise, at this enforced separation but wise enough to offer little resistance to a beast so inflamed and so much in the grip of his passions as to brook no argument whatever. For the duration of the rut the displaced knobbers will usually linger in the vicinity of the herds, despondently, sometimes in small groups, at others, dotted about singly, every so often attempting to sneak back to

Roaring, against a background of shattered rock.

Even an unseasonably early fall of snow in October does not cool the hot-blooded stags.

A disconsolate 'knobber,' evicted by the 'master' stag at the rut, lying away from his mother and the herd probably for the first time in his life.

the family group, and equally often roughly and rudely charged out again. Here and there a young antlered stag may also linger about in the vicinity of the hinds, old enough to feel desire but not strong enough to do much except to look and long. Not that this keeps the more ambitious of these young stags from attempting periodic forays into the hind groups, and the 'master' stag is kept perpetually on the 'qui vive', running out the juniors and at the same time ever alert for older and more serious competitors.

Young stags, much as they may aspire to the acquisition of hinds, have no chance when the master stags hold sway.

A 'warning-off' roar from a stag with hinds as he hears a distant roar from a possible rival.

During the rut a stag holding hinds, prey to his own fears, is ever watchful.

Though the stag is so unreasonably, to us, in thrall to his passions at this time as to seem literally to lose all appetite (his normally full stomach containing, as the rut progresses, only an evil-smelling, evil-looking, black semi-liquid mess which sloshes about in a space three-quarters empty), I am convinced that the other most telling contributory factor in his inevitable loss of condition, weight and, eventually, energy, is this almost incessant running and racing to and fro from one fringe of the herd to the other, and *not* the actual serving of whatever hinds he may be holding. A herd of thirty or forty may look over-large for any stag but this herd will contain a larger proportion of beasts not yet sexually mature than otherwise, and even of those old enough to be sexually mature I have never been able to detect more than two, more often only one, 'ready' for the stag on any particular day.

Not only does he expend much energy in chasing out pestilential youngsters and occasionally in defence against an aspiring rival strong enough to try conclusions with him, he also wastes much energy in almost perpetual and it seems purposeless chivvying of his hinds, apparently quite unable to see them at ease while he himself is not free to relax. The

A stag holding hinds is an absolute despot.

rutting stag is indeed seldom at peace. If, worn out, towards the end of the rut he does lapse into a head-drooping involuntary doze it may be only seconds before a nearby roar or a sensed, rather than heard, movement among his hinds galvanizes him into explosive wakefulness. Even in the early stages of the rut, while he is yet travelling to where he hopes to find hinds, his heated blood has him thrashing at low-growing birch branches, leaving them dangling, broken and barkless behind him, or at clumps of rushes where trees are denied to him; or even, as I once witnessed, forking up heather and moss, tossing tufts high in the air, looking almost foolish at times with a chaplet of green sphagnum moss festooned around his brow points—but with nothing foolish about the wildly glaring eye or the tangible air of menace that hung around the whole beast, black from recent indulgence in a peat wallow.

The 'thrashing' instinct must start early in red deer males for I once watched a knobber, in October, going through all the motions at the branches of a fallen birch tree although he had no antlers, merely knobs. Wallowing takes place quite a lot during the rut, both hinds and 'travelling' stags indulging in this, but seldom does a stag actually *holding* hinds feel secure enough to take time to wallow. A stag wallowing uses a peaty pool that is semi-liquid in the centre with firmer edges, and flails and strikes at the soft wet centre with lunging foreleg, alternating this with a vigorous stirring, using his antlers, until he has his black 'pot' to a porridge-like consistency to suit his taste, whereupon he subsides into it, stretching out luxuriously on one side, plashing perhaps with a foreleg, meanwhile scrubbing head and coarse-maned neck hard on the firmer edges of his murky bath. He rises, dripping, only to subside onto his other side and repeat the process until the only sign of colour, other than peaty-black about him, is a hint of grey about forehead and nasal ridge and around the gleaming eyes. He may suddenly erupt out of his bath as if the temperature had somehow risen to scalding-point and race off, bucking and prancing like one stung by a million hornets, droplets of peat and water flying all ways, head and antlers shaking frenziedly—and then, as if all at once sobering up and recollecting the more serious aspects of the rut, he will come to a halt. A roar, one more lusty shake perhaps, then away, a jet-black stag, once more in search of hinds, and now even more intimidating-looking than ever.

Hinds wallow in much the same way but lack antlers to aid the flailing foreleg in stirring the mixture up. I have occasionally seen hinds roll completely over, momentarily on their backs, four legs kicking in the air, but never, so far, have I seen stags roll over thus.

Once or twice I have read references to stags masturbating at the time of the rut but this I just do not believe. What I *have* seen, and what may well have given rise to this theory, is, in October, a lone stag put his head down to between his forelegs and then urinate over it, waggling head and antlers slowly from side to side, luxuriating in it as a positive jet of water, gleaming silvery-clear, sprays head and neck. On occasion he may then lie down and roll in the messy odoriferous moisture he has himself created. Again, I have watched a stag serve a hind (after four unsuccessful attempts) coming bolt-upright, a huge, black exclamation mark, with a galvanic thrust which literally propelled the hind some yards away while he dropped down to all four legs again and walked stiffly across to a big clump of rushes. This he sprayed with his urine, long and liberally, thrashed it about with his antlers and then lay down and rolled and rubbed in it. No wonder a rutting stag smells so rankly by the end of the rut, his belly-hair in a curling, black-stained line, forever moist-looking.

A stag serving a hind, a sight seldom witnessed.

I believe myself that by far the majority of hinds come into season and are served in the first three weeks of October and that there are just sufficient 'late' hinds to keep the young stags, learning the trade, as it were, in a ferment until well into November. Most of the hinds must be absolutely fed-up by then, after more than three weeks of almost incessant chivvying around by perhaps a succession of stags. The in-season period of any individual hind lasts for only eighteen to twenty-four hours and she will 'come on' again about three weeks later if, as rarely happens, she is not served in her first time 'on season'. A hind, particularly a young hind, can be quite demonstrative while 'on season', soliciting, quite brazenly to human eyes, the attentions of the stag, rubbing herself along his flanks, licking his 'face' and neck and even rearing up on him. To a stag in the closing stages of the rut she can become over-importunate and she may get a savage dig or two from his ever-ready antlers, as a 'hint' to wait until *he* is ready. I have, in fact, had two reliable eye-witness accounts of hinds, which had thus aroused the ire of a stag, being killed by a particularly vicious though probably inadvertently lethal stab, for the stag is no gentle wooer, he is much more of a tyrant than even the most despotic of sultans. I quote from my notes of 10th October 1957:

'Watched a stag with a big herd of hinds of which two only seemed at all ready for his attentions. He "covered" one hind *nine times in a space of twenty minutes* without being successful in serving her, confirming my idea that he was pretty well "run" (i.e. nearly at exhaustion point with the rut). This hind was a young one, possibly in her first-ever season and she was very "keen" indeed. Twice she "mounted" the stag, at intervals she licked his face and she followed him around so continuously that once, losing patience, and perhaps sensitive to his own "shortcomings," he "forked" her peremptorily away. The "mounting" procedure of this stag invariably followed the same pattern. Each time, before rearing up, he rested his chin on the rump of the hind, was up for a moment, then slid off and as he came down and stood alongside the hind so she put her head back and licked his face before running off a few yards, wagging her short tail comically. For all of the two hours of my watch that stag was never still and, ironically enough, he paid far less attention to the two hinds in season than to roving and weaving about among the hinds, or running an occasional young interloper out. There was no doubt that the hinds not actually in season were at times in mortal terror of their "master". A recumbent hind, placidly chewing her cud, would quickly jump to her feet and run at his approach or she might

A stag with his small group of hinds in typical rugged terrain, all alerted by the nearness of the photographer.

instead stretch long neck and head forward along the ground, extended to its fullest extent, submissiveness personified, while the stag stood menacingly over her until some other movement diverted his fickle fancy.'

Little wonder that that stag was dispossessed only four days later by another of similar size that had held a much smaller herd in the vicinity for a few days and seemed a lot fresher than the one I had watched on the 10th. The dispossessed stag

Turning to run as their suspicions of the proximity of man become certainty.

attempted to make a 'comeback', as I lay watching on the 14th of October, but after a short sharp struggle he was conclusively ousted a second time. He hung about until evening before making off, finally, it seemed, accepting the inevitable.

With a fresh and vigorous stag a hind seldom needs to be too importunate for he will pursue her until he serves her. The recipient of his rough 'wooing' will stand still for some seconds after having been served, head and ears drooping as if in utter dejection, yet paradoxically, her short tail will be wagging like mad, and then she will 'revive' and lose herself in the anonymity of the other hinds. Undetectable she may then become to our blunted human senses but not so to the stag which may yet seek her out again while she remains 'on season' and, if his herd is disturbed and consequently split up by human or other agency, will do his utmost to retain her, even if he loses the others of his herd. On one October day in attempting to photograph a nine-pointer in the act of serving a hind I disturbed just such a herd and away they all went towards another herd about a mile farther up the glen, held by a strong eleven-pointer. The nine-pointer was slightly smaller than the eleven-pointer and without even a pretence at a struggle the eleven-pointer took over all the nine-pointer's hinds, *except* the hind in season and her calf. These the nine-pointer retained, driving them adroitly past the eleven-pointer without halting in his headlong retreat, what time the 'eleven' was fully engaged in rounding up and adding to his original herd the bulk of the herd that fate, in the guise of myself, had flung in his way.

It is when adjacent herds are disturbed and then consequently intermingle that much of the fighting between stags witnessed by humans, themselves the cause of the disturbance, takes place, more often than not in a brief flurry of cervine violence. Sometimes, though, they are long and thrilling enough to rivet the human watchers in their concealed viewpoint, enthralled by the cat-like agility of the contestants as, having first fenced about, head on, with their antlers until they get their foreheads into contact, they then wheel and lunge, pushing with all the strength and endurance fostered by a life on ground at times seemingly more suited for mountain goats than red deer, so steep, rocky and inhospitable can it be.

Fights take place for possession of the coveted hinds, the contestants first sparring about to clear each other's antlers.

Antlers play no part in this Titanic shoving-match, as is demonstrated by the fact that a heavy enough hummel (antlerless stag), if also endowed with sufficient strength and agility, will *at times* defeat a stag possessing the outward advantage of antlers but not the less obvious advantage of weight and strength. In stags equally matched in weight and strength, though perhaps differing widely in their antlers, agility and also sheer tenacity will probably determine the eventual winner of what is invariably, in cases like this, a lengthy struggle. Ground advantage, i.e. being uphill of one's opponent, counts for a lot in the shoving-match of the stags' battlings, and rarely is a battle fought out on level ground in the Highlands. The hill-slopes being there, each contestant is

Antlers cleared, foreheads come into contact.

continually striving for a position uphill of the other so that he can shove his opponent back in a peat-slicing, hoof-plunging, seemingly irresistible rush, until the one being shoved back, in a single lightning wheel suddenly succeeds in reversing the positions and becomes in turn the one with ground advantage. So concentrated and intense can such an equally balanced struggle be that the combatants are aware of nothing but the absolute *need* to triumph; indeed, where distraction may mean defeat, they cannot heed even the thing above all else capable of distracting them, the presence of man.

Most struggles between stags are indeed brief, for the odds against two perfectly matched rivals meeting are high; but twice in one season I was unbelievably lucky in witnessing fights of this nature, one of which ended in stalemate, with two utterly exhausted stags parting as if by mutual consent. The second one I shall never forget and probably shall never see the like of again; it raged up and down a stretch of hillside while I watched; superlatives cannot express the speed, agility and die-hard tenacity of the two involved. At first I had lain hidden, having first wriggled to my view-point of that epic battle, but before it ended, so obviously blind to everything else were the stags, I was *sitting* in plain view about twenty yards from the engrossed fighters. One and then the other seemed to be winning and if a few bare seconds before it finished, I had been asked to forecast the winner I should have been wrong, even at that late stage. It was then that my money would have gone on the stag who seemed to draw on some hitherto untapped reserve of strength and in one headlong, precipitate, mad rush pushed his opponent right across a small flat and *up* the slope beyond. Spectacular and heart-stirring as it was, it was a tactical error, for the other, being pushed *uphill,* mustered sufficient strength to halt his opponent's apparently victorious 'shove-back' and hold him, both stags graven statues as nobly proportioned as anything from the Parthenon, for an interminable second, before pushing him in an accelerating, plunging downhill retreat. Without warning hint to the human eye, the stag now being pushed suddenly pulled back, disengaging antlers in a brief clatter, and ran for life itself; behind him, without any pause for self-laudatory roar, raced his conqueror, finding time, swiftly as they moved over broken terrain, to swipe and fork viciously at the part of the vanquished stag nearest to him, his fast-receding haunches. If the loser had stumbled or slackened pace, there is no doubt that he would have been severely stabbed by the lethal brow points, but he had obviously retained enough strength to make a sure-footed and fast retreat and he escaped uninjured, in body if not in spirit.

Seldom, actually, do these contests end fatally though a fairly common serious injury is the loss of an eye, more often than not, I suspect, by an accidental, rather than intentional, thrust, possibly in the preliminary sparring to clear each other's antlers. Superficial injuries such as torn ears (ragged-edged thereafter for life), antler-gashed forehead skin and stab-wounds in ribs or haunches (easy to visualize in terms of the battle just described) occur commonly enough also, of course. On examining a stag with very poor antlers shot in one season I was amazed to find the broken-off tip of an antler point, scarcely more than an inch long and about the thickness of one's little finger, embedded so as to be barely perceptible in the thick skin of his neck. It left me marvelling at the venomous strength behind that well-aimed thrust, at the fact that the antler point had actually been broken in the act, and at the toughness of the neck skin which had saved the recipient's life in that particular battle.

No conventions inhibit the withdrawal and speedy retreat of the beaten stag.

Though, in my experience, fatalities are few in fighting between stags it is incontestable that antlers *are* capable of killing and that the desire *is* there in the case of rut-inflamed stags. Once and once only however have I found the carcase of a stag that I am sure was killed by a direct antler thrust. Right on top of the skull, centrally placed as with precision, just forward of a median line between the antler coronets, was an irregular, jagged-edged hole which had obviously penetrated the brain. It was much too large and irregular for the entry hole of a bullet and in any case there was no exit hole.

Incidentally, it is the *brow* points of a stag (the first long points above the forehead) which are the lethal ones, not the top points at the extremities of the antlers, so that, in my opinion, a 'switch', or spike-topped, stag is potentially no more dangerous to his fellows than the more desirable types with forks or cups on antler tops, despite theories to the contrary. There is also no chivalry whatever in a red stag's

Seldom do fights end fatally, but one eye blinded by a chance antler thrust, as shown here, is a not uncommon injury.

make-up. At the rut, or indeed at any other time, he is *completely and utterly selfish,* unwilling though one is to have to admit this in a beast one admires so much otherwise. In red deer, to paraphrase a well enough known saying 'the female of the species is more selfless than the male', for a red deer hind with a young calf is more than admirable in her selflessness at times. Not so the stag; 'finer' feelings seem entirely lacking in his make-up and any weakness in another stag, at the rut particularly but also in competition for grazing or shelter, is immediately taken advantage of. I know of three *definite* instances where a stag only seconds from death after being shot in the stalking season was attacked, viciously and remorselessly, by another stag seemingly materializing from nowhere, with no hinds at issue and nothing else apparent at stake, simply an expression of irrepressible spleen and savagery vented on an 'adversary' whose weakness and inability to resist were immediately realized and as immediately taken advantage of. It will be sufficient to describe one such instance: in this case the presence of the second stag was known before the first was shot.

After a long, hard stalk a very good and reliable friend of mine was just getting ready for its culmination in a shot at the very poorly antlered stag he had painstakingly worked his way in to, when, over the ridge beyond 'his' stag, came a single

This stag narrowly escaped death in fighting; he had the broken-off point of an antler imbedded in the thick neck skin.

'travelling' (i.e. looking for hinds) stag, roaring wildly. The original stag advanced some way from his hinds to meet him, and the customary stiff-legged pacing, side by side, only a foot or two apart, began in a wide arc around the hinds. One's first conclusion is that this invariable procedure in such a case is completely ritualistic, irresistibly instinctive in asserting itself. Then, as one notices that the 'defending' stag always keeps on the *inner* side in this round of 'his' herd, so that to have any chance at all of getting at the hinds the interloper has first to try to best their 'possessor', one realizes that it is entirely logical. The encounter may not even come to a fight, for although at any unpredictable point in this pacing both stags, as if activated by one single steel spring, may wheel and engage each other head to head after antler-clashing preliminaries, equally unpredictably the challenger, having weighed the opposition up, may, after a 'face-saving' period of pacing interspersed with a venting of his disappointment and frustration in a vindictive thrashing at the inoffensive heather, turn and stalk off. No pursuit is offered by the defending stag; one almost envisages his wiping drops of sweat off his brow as he returns to his hinds and, still supreme, stands and roars his relief at his retreating erstwhile rival. To return to my account, however: in this instance the issue *was* decided by combat. The stag originally stalked was worsted and while the winner raced to take over the hinds, he began to make off and was then shot. No sooner did he drop than the other stag, ignoring alike the hinds he had just acquired and the sound of the shot, ran across and viciously set about the fallen stag. This so incensed my friend (illogical as you may feel this was) that, his sense of fitness of things sorely tried, he promptly dropped the second stag literally on the body of the first, although he had intended to leave the second one 'in possession'.

Among the things we still speculate over, in red deer, is what determines *which* stag going *where* to rut. There are estates which regularly feed their stags in winter and in doing so get to know well their 'master' stags by shape of antlers, number of points, width of span, etc. Some of these 'master' stags they will see, season after season, rutting on their 'own' ground; others will vanish at the onset of the rut, and at times even the stalkers of adjacent estates, who may be 'briefed' on the recognition of a particular stag, will be sure he does not rut with them. Yet when winter feeding begins again, these 'far-travelling' stags will be back again. I know *definitely* of two widely contrasting cases involving stags that had been ear-marked as calves. One I had tagged myself; it was shot, at the age of six, on an estate at least *twenty miles,* as the crow flies, from where I tagged him as a recently born calf. The other was shot, as was the first, rutting and with hinds, also at the age of six, but no more than *a few hundred yards* from where he was ear-tagged. Where does one go from there? My own tentative opinion is that a stag will wander maybe many miles each rutting season while he is yet too young (aged three to five, say) to hold hinds, each time returning to his wintering area, but on attaining sufficient years to have success in acquiring hinds he will 'travel' only the necessary distance to where he can oust a weaker stag, which may be many miles or a matter of yards. If on attaining hinds he is strong enough to hold them for most of the rut he will almost certainly return there year after year at the rut, not only to the same estate but to the same coire or glen on that estate. This has been demonstrated to me as I have recognized stags year after year rutting thus, so that one comes to expect them and feels a pang when inevitably one year they do not reappear.

And yet, in contrast again, season after season one sees 'unknown' stags, not, as is logical, confined to 'up and coming' maturing stags, but often stags obviously more than just mature, full of prime and vigour. Such stags must definitely have held hinds elsewhere in previous years. Why then their appearance on 'strange' ground? Ousted by some positive Hercules of a stag and consequently forced to roam at large again? Or simply wanderlust? I do not know, though I suspect the former to be the more likely. In connection with this I remember vividly one such demonstrably masterful stag appearing one season, an unmistakable stranger because he was a *big* hummel. A true hummel is permanently antlerless because he has no pedicles—the circular (in section) bony 'roots' from which the antlers grow annually—but simply rudimentary 'bumps' of bone on an otherwise smooth skull. We were stalking a stag holding hinds at the time and were just about to advance across some broken ground which temporarily hid our deer from us. The stag was a nine-pointer of poor shape and no symmetry, an undesirable in fact.

A big hummel, an antlerless stag with only hair-covered knobs where the pedicles should have formed.

Suddenly, as we had half risen, so did a veritable voice of brass assail our ears, dropping us flat again simultaneously as if felled by it. It came from just over a ridge to our right and the roar of the nine-pointer, impressive enough before, now seemed as innocuous as the weak mewing of a new-born kitten in comparison with that masterful bellow. It remains to this day the harshest, most compelling roar I have ever heard, with an awe-inspiring rasping timbre to it, even to a 'hardened' stalker.

'What kind of stag is this?' was, I'm sure, a mutual, though unvoiced, query and almost at once we were vouchsafed the answer: a big hummel, which answer caused us to shrink even flatter into the heather which now felt all too short. Awesome he looked, even at the distance he was from us, breasting the evening skyline, in violation of its serenity. Black he was with peat, hirsute and matted of mane, but with the oddly repulsive antlerless grey poll of the big hummel. Coming our way indeed he was, but so avidly, vociferously and rabidly intent on the hinds of the nine-pointer, now almost forgotten by us, that he passed us by, at speed, without suspecting our presence. Out of our sight he went in the direction of 'our' hidden deer. With one thought only in our minds, to see the surely inevitable combat, we were about to rise again when, so fast was this master of the 'coup d'état', that a terrific outburst of roaring held us spellbound again, culminating in the sudden appearance of 'our' nine-pointer, in speedy and unashamed flight.

Rarely has excitement gripped me so absolutely as in the ensuing crouching, fast but cautious rush across the broken ground to see what 'directions' the hummel had imposed on his easily won herd. Were they indeed still there, or had he driven them on in impetuous rush? They *were* still there and even as our eardrums were again nearly riven by that tremendous voice so did the rifle's 'crack' end the hummel's short reign, and darkness descended on him at the zenith of his victory. He proved to be about eleven years old; another short year or two and, with the waning of his powers, he would have known the accelerating occurrences and bitterness of defeat himself, for rarely does a Highland stag of more than twelve or so retain enough physical strength to have the edge on younger and more vigorous opponents. Sixteen stones was his weight and the nine-pointer, poltroon though he appeared on that memorable occasion, proved to be a full two stones lighter when, two days later, back with the same lot of hinds, he was brought to book. Giving away two stones made his speedy defeat easily understandable.

When a master stag has held hinds for about three weeks he will have 'had his fill' and may then voluntarily abdicate. Typical of this was the case of a ten-pointer I knew well for five successive seasons. On 22nd October one year he left his hinds, after holding the herd, sometimes adding to it, sometimes losing a few, but always retaining a nucleus, from 5th October. At 9 a.m. on the 22nd I watched him lying quietly, all alone, drowsy-looking, some way from his hinds, now stag-less, and I left them thus as I carried on out farther.

After holding hinds for about three weeks a big master stag such as this may have had enough and will quietly 'abdicate,' journeying slowly back to where he will winter, in amity with his fellows once more.

On my way home, about 4 p.m. I again saw him, this time miles farther out, leisurely, but with an implicit air of purpose, making his way over a skylined ridge, away to his wintering ground, passion appeased for that year.

The end of October sees most of the big stags away and it is then that the young stags 'come into their own', striving (often two, three, even half a dozen, among the one herd of hinds) in competition for the few late hinds still to come 'on season'. I once watched, in the closing stages of a season, a young stag pursue a young hind over more than a mile of hillside, turning and twisting after her, nose thrust forward and antlers 'lying back' on his shoulders, through two or three herds of hinds with young stags, these always tagging on in pursuit momentarily but always giving up to rejoin their 'own' hinds, so that at times the hind was being avidly pursued by two young stags with outstretched lecherous-looking profiles, lips wrinkled back in grotesque leering eagerness. Eventually, having described a wide circle the hind rejoined her group, the staggie still hard behind her, and no doubt he was 'rewarded' for his persistence later, though this I could not wait to witness.

And so the rut tails off, often the first early snows coming while yet an adolescent roar or two is still to be heard. 'Peace at last', one imagines the hinds feeling, no worries as yet of the winter to come, only relief at the cessation of the constant chivvying of the rampant stags.

Only after the peak of the rut is over, and the master stags depart, do the younger stags have a chance with the hinds.

Winter SNOW, FROST AND STORM

With the ending of the rut and, with it, the despotic rule of the mature stags, the hind herds, the 'displaced' knobbers now back among them, drift gradually out to the higher ground, with its grazing still plentiful enough, until the first severe weather of winter brings them to the shelter and *slightly* kinder clime of the middle ground of the hill, as distinct from the lower glens which will then more often hold stags than the hardier—or less bold—hinds. There are times when we get heavy snow in November (even in October, at the rutting time

After the rut is over the hinds usually keep to higher ground than the stags.

The wintering stags are usually to be found availing themselves of the better grazing and shelter of the lower glens.

in one year, we had a foot of snow, against the whiteness of which the roaring stags seemed incongruous indeed), but these early snows are usually short-lived and the hinds are still fit and well-fleshed after the good grazing of summer and autumn, and well able to cope with the bad weather. The stags, the mature ones more particularly, are in poorer condition after their 'excesses' at the rut, as a result of which they may have lost one to two stones in weight. The ability of red deer to cope with conditions of deep, soft, cold snow has often left me marvelling, and envious. Time after time I have been treated to the sight of their apparent disregard for the fierce squalls of winter's blizzards and the way in which they

Hinds appear to be hardier (or less bold) than the stags and stay out on higher ground even in snow.

can extract the maximum shelter from the minimum afforded by hills devoid of trees. Their main regard is to find some form of windbreak and they are skilled in this from an early age. In this respect their hill habitat, with its innumerable ridges, hollows and areas of ground broken up by peat hags, is of some advantage when weighed, for instance, against the flatter tundra of parts of the Arctic. The herd lies, some chewing the cud, some just lying relaxed, others actually dozing, each beast in the oval 'bed' in the deep snow created by its own body's weight and warmth, with every appearance of comfort, even when a raging snowstorm converts them, temporarily, into 'white deer.' Fortuitous as this camouflage is, it can be utterly effective. Especially before I learnt better, pushing on fast into the hill after sheltering myself temporarily from such a snowstorm, I have watched, incredulously and impotently, cursing myself for my blundering, an apparently empty white hillface transformed before my eyes into a herd of fleeing deer, losing their white coating in the suddenness of their flight.

December, in the north-west Highlands at any rate, is on the whole a sombre month, a month of dismal greys and short daylight hours. Traditionally a snowy month, of late years it has been more often a wet month, days of cold, drizzle or, much worse and more strength-sapping than snow, of

The stags look much darker in their winter coats than in the red of summer.

In early winter the hind's thick dark coat is sleek and elegant.

Red deer will lie out in a blizzard and become almost as white as their background.

The calves still have an unmistakable juvenile look although their dapples have long been replaced by their first thick winter coat.

merciless, chilling, horizontally-driving spear-points of 'hard' rain. Against even this penetrating wetness, however, the winter coat of the deer affords superlative protection, evolved of absolute necessity if the species was to survive in the Highlands. Quite distinct from the *single* red coat of summer, the winter coat is a thick *double* coat, the wiry long guard-hairs of the outer coat resisting both rain and snow and

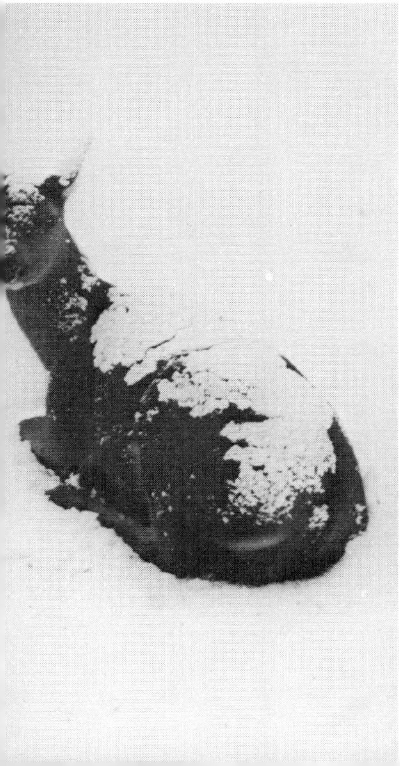

keeping dry an inner coat of a warm 'woolliness' akin to rabbit fur, blue-grey in shade, which insulates its wearer from the cold blast of winter. Red deer may lie for long periods of rain or snow, getting apparently soaked, but it is only the outer coat which gets wet. When the rain or sleet ceases, beast after beast rises and, with a vigorous shake, surrounds itself for a second or so with an aura of silver-grey 'mist', before subsiding again, or resuming grazing, with a coat now, to outward appearance, quite dry.

The first of the heavy snowfalls must be a strange, indeed inexplicable, experience to the calves of that year, obscuring their everyday world with a soft, fluffy, tangible substance which yet melts to nothingness if they essay a tentative tasting of it. It is then that parental experience and example are of great value to the youngsters, teaching them to 'scrape' down, with that beautifully fluid semicircular sweep of a foreleg, to the grazing below the snow's softness or to browse, in the wooded fringes of the lower glens, on twigs within reach. Red deer browse more than is perhaps generally known, and will readily cut, and gradually chew in, twigs of up to at least $\frac{3}{16}$ inch diameter, using their strong cheek teeth in cutting and chewing, the twig aslant in their jaws. The eight sharp-edged front teeth they use only in grazing. These they have in the lower jaw only, and they cut upwards against the hard gum

A sweeping flailing movement with a foreleg clears the way down through soft snow to the grazing beneath.

The deer graze on the patch thus cleared until it is 'eaten out.'

pad of the upper jaw. The cheek teeth are the chewing and cudding teeth and these the deer have six to each side in both upper and lower jaw, thus giving strength in cutting, while browsing, as distinct from the plucking of the normal 'soft' grazing for which the front teeth are used, grazing which forms the bulk of their food intake. For browsing, some species of trees seem more to the taste of deer than others; birch, rowan and, in some of the river banks, ivy are the species most often available but, curiously enough, elm and gean (wild cherry) are favourites where available; the grey-green lichen which bewhiskers the older tree-trunks is also eaten. All these are of value and most often resorted to when the grazing is covered by snow.

I have noticed that red deer, herd animals anyway, in prolonged periods of deep snow, tend to gather into larger herds than are usual, herds which are adjacent to each other amalgamating, as it were. This again probably has its value. Where a large number of deer are scraping and making tracks and 'passage-ways' through deep snow the stronger are, willy-nilly, 'breaking out' the going and exposing grazing for the younger and the weaker members of the herd.

Having to endure, as they do, weather of almost indescribable foulness and savagery, from as early as October until as late as the end of April in the year following, the hardiness and survival rate of Highland red deer are truly admirable, facts to be remembered when their slighter weight and smaller antler development are contrasted, to their disadvantage, with those of Continental red deer.

Even in these long hard months, however, we can have days of bright sun, the more valued for their scarcity, and these days the deer only too obviously enjoy immensely, having dry backs and beds and at least an illusion of warmth from the sun. Their reaction then is far different from their behaviour in June to September, when the sun is strong enough to make them seek for shade. In the colder months every opportunity to enjoy the vitalizing rays is taken advantage of. One particular incident I have written of in an earlier book but I feel it will bear repeating here, this being a book on red deer behaviour. One December afternoon of bright, wintry sunshine I came, unobserved, over a concealing knoll to see, below me, a red deer hind with a rather puny-looking calf grazing, the hind herself an obviously old one. I managed to get down into the heather before being seen and, taking my spying-glass out, began to watch them. Hardly had I settled myself thus than the hind ceased grazing, went over to her calf and began to lick its neck, her long pink tongue, rhythmically busy, plainly visible through my glass. While she did this her eyes were almost closed and her calf held its muzzle pointing slightly upwards, stiffly, as in a trance. It was abundantly apparent that both were deriving much pleasure from the licking for the two minutes or so for which it lasted. When the hind ceased she simply stood, lazily chewing her cud, just as obviously enjoying the faint warmth which the rays of the December sun held. As her jaws moved in her regular chewing, so good was the light and so close was I, that I could see that she was badly gap-toothed in her incisors and I could see also the lump of cud travel up and down the 'food-tube' of her long thin neck, as she periodically swallowed and renewed this. Definitely an old, 'getting done' hind, near the end of her life-span with a calf which might well be one of the mortality victims in the following spring, yet, without fore-knowledge or thought of this, extracting the maximum from the good conditions of that day and in doing so giving me an unforgettable little cameo of red deer hind and calf.

In another winter I had a striking example of the strength of maternal feeling in a red deer hind. I was walking parallel to, and above, a sheep fence (i.e. not deer fence) which divided a low-ground glen from the 'middle' ground, when I spotted a ragged-appearing hind feeding below the fence. Close by her, her 'follower' (that is, her offspring of the previous year, about nineteen months old) was also grazing, a hind also, but smaller and darker coated. The ragged hind and her follower saw me as quickly almost as I saw them, but stood their ground, inexplicably, until I was quite near. As I came to where I could see a section of the fence hitherto out of sight I understood, belatedly, why. Trapped by a hind leg in the fence, by the crossing of the two top wires, was her six-month-old calf. As I went to free the struggling calf the others disappeared, albeit reluctantly, around a nearby ridge. Her calf had obviously been trapped for a day at least, for the leg, just above the hoof, was grooved to the bone and rust-marked from the tight wires. The calf was quite lively considering the length of time she had been 'trapped', but the leg which had been caught was very stiff from the enforced restriction, and its lower bony part between the mid-joint and the hoof seemed frozen. I worked at the leg for a while, bending and straightening it and rubbing the apparently frozen bit vigorously before I left her, lying as comfortably as I could contrive, in a thick 'bed' of heather.

Continuing my way out to the hill beyond, I met the mother head-on, almost bumping into her, as I rounded the low ridge above the fence. Anxiety for her calf over-riding all apprehension of humans, she was returning into the very jaws of danger. My sudden alarming appearance made her retreat again of course but I have no doubt that she did not go far and soon rejoined her calf. One wonders what her feelings about it all were, the quite inexplicable inability of her calf to leave the fence it had just jumped and join her, a problem she had 'solved', temporarily, by herself remaining and grazing nearby. Then after a lengthy vigil the alarm-creating arrival of a man, making her run, yet despite her understandable apprehensions, she had obviously remained nearby with every intention of rejoining her calf. Thank goodness I *had* been out in that particular area of the ten thousand acres I had choice of, otherwise an even lengthier vigil would have had the inevitable ending of a dead calf, culmination of an all too slow starvation.

January usually brings more severe weather than December and a further month has gone by during which the grazing left over from the growth of summer has progressively been reduced. Despite this, throughout January there is usually enough to keep the deer in 'good heart'. Just as well, as the following two quotations from works on Scotland in winter will emphasize: 'One cannot frequent hills and avoid storms', and 'The North of Scotland is considerably closer to the Arctic

Puny as they may look against the immensity of snow-clad hills, red deer are usually well able to cope with even a Highland winter.

Circle than are other mountainous parts of the country. In winter, conditions can be more than Alpine, they may be arctic.'

Heavy snow, mind you, has its own compensations for those with any eye for beauty, transforming the winter drabness of the hill. A delicately poised lacework of virgin snow coats every tree in the sheltered glens, even to the slenderest of their twigs, an eye-dazzling white. One has to be out early after an overnight fall of snow to witness this, for even the slightest hint of a breeze will result in a cascade of displaced snow from every twig, but it is well worth an early rise, a white world too breathtakingly beautiful to render adequately into words, scintillating with a million sparkles under frosty sunlight shining from the deep chilly blue of a cloudless sky. Even the rockiest and most forbidding of hills are softened by their white blanket, an utterly illusory softness, of course, but rewarding to the human onlooker who does not have to live out and find food in it and who has the certainty of a cosy house, warm fire and dry clothing to return to. Deer, in stark contrast, do not even have the relative comfort of rock hole, den, or sandy burrow to shelter in (even that other denizen of the hills, the wild goat, uses rock caves in bad weather, where available), but simply the lee side of a ridge, or, in extremity, the sparse birch woodland of some of the more favourably situated hill glens.

On such a day tracks of deer radiate outwards from where they have spent the darkest hours of the night, each cleft track shadowed icy-blue in the deep snow, a betrayal of the deer's progress to any human follower.

A period of hoar frost can be equally beautiful and even more chillingly cold, with every blade of grass and every sprig of heather, in early morning, gemmed and encrusted with brittle silvery-white crystals. A world of transient delight and beauty in the first light of frost-clear sun, before its slight warmth gradually renders the scene mundane once more, but a world, at this stage, of frozen grazing, every mouthful freezingly cold. Everywhere and everything is frozen, the ground, the usually soft bogs, even the black lochans grey under ice, in iron-hard grip, and one's fingers, if ungloved to use a camera, stick disconcertingly to its metal-work. During the daylight hours the bright sun, even if weak in terms of

A herd of hinds and calves making their way from the cold shade of the glen bottom into the brightness of the sunlit hill above.

warmth, does help; during the night the sparkling diamonds of myriads of stars in a velvet-blue sky only serve to underline the severity of the biting cold. In the grey light of winter's dawning as the deer rise from their cud-chewing, after lying with no *obvious* discomfort on the bleak hardness of frosted hill, they will leave an oval, dark, thawed-out patch, their 'bed', in the whiteness of the rime. The long outer hairs on back, neck and belly may be spiked and stiff with the frost; this on a living, breathing animal, is testimony equally to their hardihood and to the severity of the cold.

At times, in weather like this, the lower glens may lie clothed under an almost tangible woolly thickness of frost mist, much whiter than 'normal' hill mist, out of which one

A maze of tracks betrays the movements of deer in deep snow.

The 'poetry of motion' illustrated even on snow-covered rough ground.

may climb into a world of bright sunshine, the white mist now billowing below, reflecting back the sun's rays. Below, audible but concealed, one may hear and thus trace the progress of a deer herd in its climb out to the solace of the sunlight also, the calves calling to their mothers, uncomfortable in the chill confines of the mist and not yet as stoical in enduring this as they will eventually become, while, occasionally, a hind will answer in reassurance.

Ice, present of course every winter on bog, lochs and rivers to a greater or lesser degree, does not seem to hold any terror for deer or indeed for other animals, lacking the human knowledge that underneath lies water into which breaking ice could deposit them. Certainly both deer and fox habitually cross ice-encased water in the Highlands. In the hard frost of 1963 in the north both hinds and stags were seen to cross the Caledonian Canal in the region of Fort Augustus, on the rapidly re-frozen 'ice-floes', a jumbled, jigsaw pattern of which was left daily after the ice-breaker boat had been through. This led, however, to the death of a single stag on one day of that particular hard spell. He was watched picking his way mincingly across the treacherous surface of broken and re-frozen ice. When he was almost across, a thin bit in the ice gave way below his weight and in he went. Struggle frantically as he did the end was inevitable, with the ice-edges of his pool of freezing water splintering beneath his hooves each time he

Ice seems to hold no terrors for deer, which will cross a frozen loch, as in the background here, if they so desire.

strove to heave himself out. There was absolutely nothing the watcher could do and, mercifully, the deadly chill of the water ensured a rapid end to the struggle. That stag's body remained frozen into the ice until the thaw came.

February is still very much winter in the Highlands and with the hill growth of the previous summer now fast diminishing and with no chance of renewal, until May at the earliest, over most of the red deer country, the end of February usually ushers in the most difficult time of the red deer's year. The traditional belief is that it is in the depths of winter that most of the annual 'natural' mortality occurs in red deer, but, though deep snow and frost in winter can curtail temporarily the grazing on the hill, it is in March and April that this grazing is at an absolute minimum and this is the real hungry, testing time for red deer. The death roll in the actual winter months is small, apart from exceptional occurrences of single deer, even, at times, groups of deer, being caught in avalanching snow, a relatively rare event in the Highlands.

The harshness of our Highland winter weather, then, is, in my opinion, very well coped with by the hardiness of our red deer and if we were lucky enough to have appreciable spring growth in late March and throughout April, instead of having to wait until May, this winter severity would hardly figure in the red deer existence. As it is, however, with food supply in winter and early spring progressively if gradually diminishing

This stag badly misjudged the thickness of ice on the Caledonian Canal; its body became frozen into the ice.

as a result on the one hand of the grazing of sheep and deer and, on the other, of the sheer inevitability of climatic conditions on exposed hill ground, culminating in the lean months of the Highland March and April, a combing out of any weak deer is to be expected and, callous as this may seem, is ultimately of benefit in an environment where only the fit will thrive.

Spring THE DIFFICULT MONTHS

Spring, not winter, is the really hard time for Highland deer, especially if we have late snowfalls extending, as in one year, into early May.

Accidents, fatal accidents that is, are rare in a deer population and when they do occur it is often in the 'spring' months of March and April, which may be anything but spring-like in the Highlands. Their very infrequency tends to throw those accidents which do occur into perhaps undue prominence, exaggerated at times by the bizarre nature of some of them. The main point about them really is that these accidents take place chiefly at a time when the deer are at their lowest ebb, namely in the months mentioned above. Most of those involving deer caught in stock-fences can be directly attributed to this low physical condition, for such a fence will be jumped easily by deer in the peak of condition. In the lean months the weakened animal, in jumping, may strike the top

It is when the red deer are at their lowest ebb in condition, in March and April, that accidents occasionally occur when they jump fences they would normally clear easily. Usually the victim is trapped by a hind leg—rarely by both legs—caught between top and second wires, and the beast dies miserably unless rescued in time.

wire with a trailing hind leg, which may slip down in between the top and second wires. The falling weight and abruptly checked forward movement of the beast then cause the trapped hind leg to act as a lever which twists the slack second wire up and over the top wire, thus forming an all too effective

Loose wire or wire of rusty old fences is a potential menace to deer. The stalkers managed to cut away most of the wire caught on the antlers of this stag, which had then snared itself on a fence post. The wire cut away weighed 28 lb.

snare just above the hoof. The trapped beast pulls hard *against* the tautness of the wire, thereby tightening it, and, lacking the reasoning powers to 'think it out' and then ease its weight back to relax the grip of the wire, eventually dies in the self-created trap, unless a human comes along in time to save it. Since there are miles of stock-fences (or sheep-fences if you like) on most Highland estates this rescue act rarely happens. In many years of being on the hill I have managed to save only two beasts caught thus, the calf already alluded to in the last chapter, and a hind, which was very weak, too apathetic to struggle as I attempted to free her. When I had got her out I pulled her unresisting weight a little way to some dry heather and there set her in a lying position, actually tucking her legs under her. She tried to rise, found she was unable to, and then just lay resting for about ten minutes; then she again attempted to rise and, this time, succeeded. While I watched from a distance she stood quite still for a further few minutes before walking slowly away, shakily at first, but getting steadier on the legs she had been unable to use for so long. I think she survived, I certainly never found her body, though I made a thorough search of the area at a later date. Only once have I seen a deer caught by both hind legs, in April 1957, a hind, and she had long been dead when I found her, the body part-eaten by fox, raven or eagle, all of which eat carrion readily in the lean months when mortality in deer and hill sheep occurs.

It is remarkable how often deer will take advantage of any gap in a fence or, on peat-hag broken ground, use a peat 'runner' which goes below the bottom wire, progressively enlarging it, so that such a 'runner' becomes eventually an accepted 'pass', which the deer will use rather than jump the fence, as they could easily do. Even the biggest of stags do not consider it beneath their dignity to 'crawl' under a fence thus, adroitly ducking head and antlers under first, and then 'creeping' below, where they could readily and much more elegantly have jumped. Some of these 'passageways' are incredibly small and it is enlightening to discover that, high as red deer stand, and wide and apparently cumbersome as the antlers of the stags may be, they *can* adapt to 'creeping' through a space which at first glance seems impossible, and can acquire the knowledge to turn, twist and duck their antlered heads in going through such a space in the certainty that their bodies will then 'follow through'.

One hears the opinion, indeed at times an emphatic statement, expressed that hill deer, with instinctive or acquired knowledge of their terrain, *never* bog themselves. It is probably in the main true that deer seldom make fools of themselves, in their necessary daily comings and goings on the hill, by blundering into 'soft' ground and becoming bogged. I did once, however, find a stag, dead, which had perished through selecting a boggy peat-pool to wallow in that was too *constricted* and deep. A mature stag, he had obviously not long cast his antlers and had probably been feeling itchy, with his 'old', shabby winter coat, tufty and loose, and had chosen the red deer panacea for this (and other 'ills' right throughout the year, from the itch of emerging warble-fly grubs, to the discomfort of summer heat and the 'hotness' of blood at the rut), a 'bath' in a peat wallow. The depth of this particular stag's ill-chosen wallow gave him no purchase by which to heave himself out and, as he sank, its narrowness had literally jammed him so that he had died miserably, caught thus. Peat-wallows being used so often by so many deer, this once-in-a-lifetime finding of a stag trapped while wallowing is remarkable in being so rare, rather than in revealing red deer to be capable of such misjudgment. Perhaps the most noteworthy use of a 'wallow' I have observed, and one that will always have its own pathos for me, was when I saw a stag with a recently wounded, broken foreleg standing in a peat hag actually nosing peat up and around the wound. I would hesitate to hazard a theory that this stag sought any healing property in the peat 'plaster' he was using; I believe he was simply availing himself of the soothing qualities of the cool peatiness familiar in alleviating itchiness, heat, etc., to which the hot discomfort in his leg was probably akin.

I came upon another unusual and probably 'once-in-a-lifetime' accident many years ago in spring on the bank of a river whose course ran in the bottom of a wooded glen, a course that it had in many places carved down, over the

The hind that had 'hanged herself' by getting her head jammed between twin birch trunks above a river gorge.

centuries, to run at the bottom of steep, rock-sided gorges. On the edge of just such a gorge two slender birch trees grew, close together and forming a narrowing 'V' at their moss-grown base. Caught between them by the head, and hanging over the edge of the gorge, suspended lifeless above the foaming waters of the river below, was a hind. She had obviously lost her footing while her head was high enough to let her reach through the wider part of the 'V' formed by the twin trees, either in search of lichen or in 'scratching' head and neck up and down between them. In any case, her falling weight had pulled her head down into the narrow base of the 'V', and had held her ruthlessly fast there. Unable to regain or use her legs, her head jammed in a vice-like grip by her own weight, she had died, fairly quickly I prefer to think.

Having written of these deaths to deer through accidents, I should like to emphasize that, distressing as every individual case must be to any human being of any sensibility, the annual toll exacted by them is *very, very slight* in relation to the number of deer in the Highland population.

There is an annual mortality among red deer, usually when the grazing is at a minimum, from late February to late April. Weak calves such as the two in this picture may not survive this lean period.

There is, however, also an annual natural springtime mortality among red deer in the Highlands, be the preceding winter mild or savage. In point of fact a 'mild' wet winter, followed by a squally and wet spring can exact greater toll than a 'hard' winter with a lot of frost, perishingly cold and icy to us, but less sapping to the strength of the cold-resistant deer than perpetual wet, above and underfoot. In two successive winters of hard, dry frost in the early 1960s, in my annual search for natural mortality I found that the loss was less than one third of that of the preceding and following milder (but wetter) winters. It should again be stressed that this annual mortality is closely related to the survival of the species *as a whole*. In nature there is room for neither sentiment nor the survival of individual weaklings; the fittest live on in order to propagate a 'fit' species, an age-old axiom but one as true today as when life began. The toll begins in late February, is increasingly apparent throughout March and continues well into April. A very few 'borderline' cases scrape through these critical months to succumb in May, or even as late as early June, the new growth then being too late to save them, worn out and weak as they are. Mortality occurs chiefly in two distinct age groups (logically so): in the 'unfit' of the nine-to-ten months old calves and the aged adult deer. In both cases wise management of deer ground, the selection of the weaker, less-thriving deer *as a priority* when stalking both stags and hinds, instead of selecting the *best* (as was once done), and regulation of numbers so that there is adequate grazing for the deer and any other grazers on the hill, can do nothing but good. Deer-stalking, once regarded only as 'sport', is now increasingly regarded as the utilization of a natural asset in hill country which has few such assets except perhaps its scenery, native inhabitants and its wildlife, and wisdom in management, rather than selfish taking out of the best beasts each season, has come to be regarded as the better policy. In this context it should be remembered that *adult* red deer in this country have no natural predators and that they have an annual natural increase which *must* be regulated. The most humane way of doing this remains deer-stalking when practised by those with sufficient knowledge, skill, responsibility and sensibility. It should also be realized that the best professional deer-stalkers have an almost fanatical

It is fairly common to find a weak calf that has died peacefully in its sleep.

regard for 'their' deer, a regard which transcends the modern obsession with set daily hours and fixed overtime rates in a five-day week.

To some of the red deer calves, then (a small proportion, I must stress, on well-managed deer ground), March or April (the latter a month in which one begins to look and long for a resurgence in plant and animal life) is the end of the trail, their first and last spring, before they have attained a full year of life. However hardened one becomes, or strives to become, it is always pathetic to come upon a deer calf dead thus. Wishful thinking as it may be, I believe most of them die peacefully, just a slipping-off of life's bonds; I have often found such calves curled up, as if asleep, and asleep I believe they were at the end, without knowledge that from this sleep there was to be no awakening. These annual deaths in the weaklings of the preceding season's calves work out, as I have already said, as roughly equal between the sexes. The mortality is invariably found in the glen bottoms and as one gets *really* to know a ground (which is only one argument for continuity of management on deer ground) one realizes that it occurs in certain areas of those glen bottoms, just as one realizes, with time, that there are areas where one is most likely to find calving hinds in June, summering herds in high

The fit hinds are strong enough to survive and bear strong calves in June.

The stronger calves, the majority on good deer forests, also survive, ensuring the continuing fitness of the herds.

summer, rutting herds in autumn, and wintering herds at the year's 'end'. My own belief is that with deer, coming downhill as they do pretty well daily into the glens to graze and shelter in the evenings, the weak ones, whether at one end or the other of the age structure, reach a stage when they are unwilling or unable to go back *uphill* with the bulk of the herd and so they linger about in the bottoms, becoming progressively weaker as they eat out the limited grazing their physical condition lets them forage for, dying at last in the most sheltered part of the glen bottom where they can find a last comfort.

By 'aged' applied to deer one means a hind or stag of about fourteen years of age, which I believe is the average life expectancy of Highland red deer under natural conditions. One finds at times adults of either sex dying naturally at twelve years or so and, rather more rarely, at an age one may estimate by tooth-wear as sixteen years or even more. In 1969 I found such an aged hind, so poor and thin as to be hardly more than a rickle of bones. I managed to take home and weigh her carcase, after estimating her age as in the region of nineteen years. Skinned out, she only weighed 49lb. The weight of an *average* hind carcase, weighed in the same way, is about 75lb., while that of a really good big hind (in the Highlands) would be from 85 to 90lb. These comparisons will give readers an idea of how thin this old hind was, a living skeleton.

A contributory cause, I suspect, to the annual mortality toll, though probably rarely a primary cause, is the infestation by internal parasites to which red deer are subject, though less subject, one must stress, than domestic stock nowadays even with their variety of drenches, inoculations, dippings and vaccines. It is well known that a healthy animal can 'live with' various parasites to which it may involuntarily act as host, but in a weak beast, the threshold of resistance lowered, parasitic infestation may build up sufficiently to become 'the last straw'. Red deer may harbour such internal parasites as stomach worms (including tapeworm), lungworm and liver fluke, and the grubs of warble-fly and of the nasal bot-fly. Calves seem particularly vulnerable; many dead calves were annually found to harbour a fairly high level of some of these parasites, more particularly lungworm, warbles and nasal bot-

Deer are subject to various parasites. Those illustrated (with an 'old' three-penny piece to give scale) are, left to right: liver fluke, warble-fly grub, nasal bot-fly grub, and lungworms.

fly grubs. Aged adults were quite often found to have a noticeable incidence of lungworm, a white, thread-like worm found in the lungs and in the windpipe of dead deer, and, at times, an incidence of nasal bot-fly and, less often, of liver fluke. There is no point in describing the life cycles of all these parasites, except to say that lungworm is possibly the most serious in weakening already weak calves, and is a parasite similar to that causing 'husk' in domestic cattle. Liver fluke affects the liver but red deer seem much better able to withstand its effects than roe deer or domestic sheep, to both of which it can prove fatal. I have never known personally of the death of a red deer directly attributable to liver-fluke.

The grubs of the warble-fly are similar to those which can affect domestic cattle. They emerge through the skin of the back of the host, in March and April, and can cause so many small holes in the hide as to ruin it utterly for commercial purposes. The most repulsive of the whole repulsive collection of parasites which affect red deer internally are, to me, the larvae, or grubs, of the nasal bot-fly *(Cephonimyia auribarbis)*. This fly is on the wing on hill ground in summer and, landing on the nose of the deer attacked, attempts to inject a drop of fluid containing live larvae into its nostrils. These larvae, minute at this stage, 'hook on', with two curved mouth hooks,

The most repulsive of the internal parasites which may afflict red deer, nasal bot-fly grubs *in situ*, near to the emergent stage, in a side section of a deer head.

A red deer hind lying at the edge of a peat wallow, which they may use in spring for alleviating the itch of emerging warble-fly grubs.

to the inside of the nasal passages and progress slowly up until they reach the pharynx, a 'pocket' at the end of the nasal passages. There they finally 'hook on' and live thereafter, feeding on the mucous of the nasal passages and gradually growing larger. By mid January they may be $\frac{3}{4}$-inch long, having a segmented body, recalling the 'Michelin man' of tyre advertisements, of flattish, oval section, rather more than $\frac{1}{8}$-inch in width, yellowish in colour with short brownish bristles on the upper side. They have what appears to be a sucker disk at their rear end and two black curved tiny hooks at the 'mouth' end. With these hooks they are so firmly attached to the host as to require quite a pull to get them off. By mid March they are approaching the emergent stage and are then about $1\frac{1}{4}$-inches long and $\frac{3}{16}$-inch wide. They eventually loosen their hold on the inside of the pharynx and are sneezed out by the host via the nostrils, or occasionally coughed out through the mouth. Occasionally also they may be swallowed and then presumably die. If the host happens to die towards the end of March the active larvae try to leave the head of the dead deer, again via the nose or mouth, and again occasionally taking a 'wrong turning' and going down the throat.

It is at this March stage when the larvae are large that I believe a heavy infestation may well contribute to the death of the host. A light infestation of a dozen or so may cause nothing more than acute irritation, but I have found over fifty, on occasion, in the head of a dead deer calf and once I found seventy-two of these large larvae in a dead male calf's head. They had formed a solid ball, utterly blocking the base of the nasal passages, defeating their own ends, and in this case it was hard not to conclude that they must have contributed considerably to the calf's death. Such very heavy infestations are fortunately rare, though twenty to thirty are common enough in deer calves in my experience and surely cannot help a weak calf in its struggle to live. The calves of the red deer *seem* to be the main host in most areas, the incidence in adult animals being slight until they get aged, when they seem more affected than in the middle age range. I believe this is because of the lack of practical experience on the part of the deer calves of the year, which, unlike the adults, take no active evasive action such as head tossing, ear-flicking, dabbing at nose with a forehoof or rubbing it violently into the heather, as already described. In the case of the aged deer, I believe their senses may be just sufficiently blunted to make them more susceptible again, the wheel having turned full circle. The mature larvae, having left the host, burrow into the soil and pupate therein for four to six weeks, after which the fly emerges and in due course the repulsive and inexplicable cycle begins again. I am not writing of this occurrence of internal parasites in deer from a scientific angle but from the angle of a field study by someone interested enough in deer to try to consider all aspects of their life in the Highlands. Do not be led into thinking that red deer are extremely susceptible and vulnerable to these parasites. As a species, they are not, and, as with accidental deaths, the actual annual toll directly attributable to parasites is insignificant, and this, may I remind my reader again, is without the aid of veterinary treatment or preventive medicines.

As a last word or two on the rather saddening but integral part of the red deer year, the annual mortality, I would add that, in nature, it is not confined to deer, it occurs in all wildlife where a proportion of young, those least fit to survive, die before they reach a year old; and, equally inevitably, if they surmount this first hurdle and eventually attain old age, the animals die when they can no longer look after themselves. There are no old folks' homes in the wild!

By late March, red deer stags have usually begun to cast their antlers, first the mature stags, and the young stags perhaps as late as mid May. On 14th May in 1973, I watched a small group of five stags. Of the five, two were obviously young stags of approximately four years of age, six-pointers, each still with his antlers on; two had obvious knobs of new growth, and the fifth was the most advanced with about eight inches of velvet-clad new antler, and brow points of about four inches in length. This will give some idea of the difference the age, and *possibly* condition, of the stags makes in their times for casting their antlers.

Why antlers are cast and regrown annually, with, presumably, a consequent strain on the bodily condition of the stags in providing calcium for this annual growth of bone, instead of being permanently retained as with horned animals (i.e. sheep, cattle and goats), to grow larger and longer as the animal ages, is still very much a subject for speculation and may well remain so. The best description I have yet read of this process is in *The Growth and Shedding of the Antler of the Deer* by Sir William Macewen, F.R.S.—a scientific work from which I here cull some basic facts:[1]

'. . . the deer's antler is deciduous [i.e. shed periodically], a new one being formed every year . . . the enormous growth of bone forming the antler is produced in three or four months . . . this growth of the antler proceeds from a single comparatively small centre of ossification situated in the frontal bone. The regeneration of the myriads of cells and the rapidity of cellular proliferation necessary for the growth of the antler is much greater than that which obtains in any other normal process of single bone formation within deer or other animal. . . .

'In the first year . . . an outgrowth . . . rises from the frontal bone which, when the outgrowth is completed, forms the pedicle. The pedicle not only springs from the frontal bone of the skull, but it is identical with it in structure, . . . having the same blood supply. The pedicle at the beginning of the growing period is highly vascular [containing vessels or ducts for conveying blood] and pulpy, and is covered with cartilage which ossifies from the base upwards, . . .The pedicle once formed is a permanent structure growing in circumference during subsequent years. . . .

'When a new antler is about to form, the whole thickness of the skull from the brain cavity outwards . . . exhibits greatly increased vascularity compared with the same parts when the growth of the antler has been completed. The blood supply of the pedicle comes from the vessels of the skull and frontal bone. The pedicle itself is highly vascular especially towards the centre from which active osseous proliferation is proceeding . . . The new growth pours out from the whole interior of the pedicle and overlaps it circumferentially, laying thereby the foundation of the future corona [coronet].

'Whenever the antler is shed, the pent-up forces which had been accumulating preparatory to the new growth find vent and from the whole upper surface of the pedicle a mass of new formation springs and overflows its edges—in form like a young mushroom projecting from its stalk.

'. . . the new antler springs from the pre-existing bone in the pedicle and skull, both of which become highly vascular preparatory to fresh ossific effort.

'. . . a thin layer of cutis [the "velvet"] spreads with great rapidity over the bone and cartilage of the antler, and keeps pace with the subsequent growth of the latter. . . .

'[The velvet] is nourished by very numerous big [blood] vessels . . . which run vertically on the antler. . . . Osseus growths are formed by

[1] Maclehose, Glasgow, 1920; pp. 1-18, 85-6.

the bone from within, growing outward between these vessels so as to leave them partially embedded in grooves. These grooves or gutters leave permanent markings on the outside of the solid antler. [Subsequently] the vessels by a somewhat similar process, become strangled by osseus encroachment at the corona and shrivel off. . . .
'[The cutis] covers the whole antler from the base to the tip not only with the epithelium of the true skin, but also with its glands, hair follicles and appendages. . . .
'Branches of the trigeminal nerve ascend in the cuticular covering in line with the blood vessels and supply the acute sensitivity of the velvet—which thus protects the growing antler during its softened state . . .
'[Preparatory to the shedding of the antler], the osseus tissue within the antler at the level of the corona becomes very dense and gradually avascular . . . until complete obliteration of the vessel ensues. The antler . . . ultimately becomes dead tissue.

About a week before an antler is cast, the skin and hair around the circumference of the pedicle, directly below the coronet, begins to show an at first barely discernible dark line, which becomes increasingly apparent (though it is never very obvious unless one can study a stag at close quarters) as the time for the antler to be cast draws nearer, as if the skin and hair covering the pedicle were, almost imperceptibly, shrinking downward. This was first drawn to my notice by the sharp eyes of my elder son, also Lea, who with Michael, my second son, shares my interest (or mania if you like) in red deer. In the case of a stag that I was able to study daily for a week before he cast his antlers I confirmed this by personal observation. The day before casting, this stag was most obviously 'off-colour' and without appetite, as also on the day he cast, when he lost one antler in the early hours of the morning in 'stirring' up a peat wallow with it, and the second one, which simply dropped off, a few hours later. I have also observed that while the pedicle increases in circumference over the years it also shortens gradually, a small portion being lost annually when the antler is cast. Antlers, dead and ready to fall, may come off in different ways though probably most just drop off by their own weight. At times the stag 'helps' them off by rubbing them on trees, or even on telegraph or power supply poles, where trees are lacking. I have found cast antlers at the base of such poles occasionally and also where stags are in the habit of jumping sheep fences. I have also been told by a reliable acquaintance, who was a hill shepherd, of

Antlers already loose at the time for casting may drop off as a stag jumps a fence.

how he once saw an antlered stag shake his head violently up and down as if in intense irritation until, suddenly, one antler was literally shaken off with such vigour that it was flung high in the air. Losing an antler apparently causes a stag a certain amount of temporary discomfort, and indeed pain, for he may run about madly for some minutes, shaking his head from side to side, until the pain, if such it is, lessens and he settles again.

Cast antlers are not readily found even when a deliberate

When the stags have cast their antlers they rear up and settle squabbles with flailing forelegs.

search is made of an area that stags are known to frequent. The type of terrain over most of the Highlands—long heather, boulder-strewn hollows, slopes of scree, peat 'drains' and burns part-hidden by heather-grown banks—and, as with the stag written of above, the possibility of a stag casting in, or at, a semi-liquid wallow contribute to the difficulty of finding cast antlers, as does also the well-known fact that red deer 'chew' antlers, bones, and an astonishing variety of other objects, presumably due to a craving for mineral content lacking in their largely peat-based grazing. Unless antlers are found fairly soon after they have been cast they are almost certain to be at least partly chewed, sometimes only the thick butt-end and coronet being left. In one spring I found the carcase of a recently dead, old stag, head and antlers intact, and on coming by it some ten days later I found that fully half of the not inconsiderable length of one antler had been chewed away and all the points of the other one had been part-chewed. I have had two eye-witness accounts of hinds chewing antlers. Nothing remarkable, you may feel, in view of this habit being well known, *except* that, in both these cases, the antlers were still crowning the head of a recumbent *live* stag. My own

Even a stag with an antler still to cast seems impelled to rear up in squabbling with an antlerless one.

Examples of antlers chewed by deer, from one with just a point or so chewed to those of which only the thick butt end and coronet are left.

personal glimpses pale into insignificance beside these. I have watched, on different occasions, a hind chew part of the leg bone of a dead sheep, most of it protruding from the side of her jaw at a jaunty cigar-like angle while she chewed with her cheek teeth; a young stag apparently enjoying his chewing at a

In the Highlands both stags and hinds will chew at cast antlers.

A most reliable stalker friend of mine watched a hind chew at the middle top point of this stag's right antler as he lay chewing the cud.

long-dead sheep's skull; and another stag chewing away, again with apparent enjoyment, at what turned out on later investigation to be the greasy and oil-impregnated leg portion of an old pair of dungarees.

Most memorable of my personal sightings of this chewing habit occurred one autumn while I was lying watching a herd of deer some 250 yards away. The deer had no suspicion that there was a human within miles yet, suddenly, a young hind left the peacefully grazing herd and came straight and purposefully downhill, nearer and yet nearer while I froze in the long heather and racked my brains as to what could possibly have aroused her suspicions. When only thirty yards away she stopped and I thought the game was really up, that she would now circle around to windward until the abhorrent human scent confirmed what appeared to be her suspicions. I could hardly have been more mistaken for, instead, she began to nose about in the heather, obviously searching for something; within minutes the mystery was solved when she picked up and adroitly manoeuvred around to between her cheek teeth three feet or so of alloy tubing, of about ¼-inch diameter, bent at right angles about half way along its length. For fifteen minutes she stood and meditatively chewed at this with every appearance of relish, an absolutely ludicrous sight and, *of course,* I had no film left in my camera. Having taken her 'mineral intake' for that day, she dropped the tubing where she had found it and returned to the undisturbed herd. It was a most interesting and enlightening sight to see her chew the alloy tubing; even more revealing was the fact that she had quite obviously known, from a previous chewing spell no doubt, the approximate position of the tubing in fairly featureless heather, and when the urge again took her had returned to find it and renew her chewing. Other objects I have known to be chewed by deer are fragments of a crashed plane, practically the entire length of a pick handle, the desiccated carcases of grouse and mountain hare, and the legs and part of the head of a 'cast' (i.e. one prematurely born dead) red deer calf, found in late April. Incidentally I have also once seen a hind licking, and apparently deriving pleasure in doing so, a telegraph pole which had been recently treated with creosote, a peculiar taste in more than one sense of the word.

With the coming of May the red deer and their *hill* background alike are looking shabby and faded. The winter coats of the deer are loose and bleached (or perhaps weather-beaten would be more apt) to a light grey in many cases and the hill growth is equally bleached with as yet no very obvious new growth. As a result of this fortuitous conjunction deer can be even more difficult than usual to pick out, at a distance, from their background. Where there are greens, especially on faces with a southerly exposure, there will the deer be, grazing almost without cessation even when, in times of better feeding, they would be lying down, relaxed, chewing their cud. Fill their bellies they undoubtedly do but it is with quantity not quality, of grazing withered and dry after a Highland winter, and a spring which is often well-nigh indistinguishable from the preceding winter months.

It will be mid May, indeed sometimes later, before there is any appreciable new growth to be seen on the hill, but as it comes and creeps higher up into the relatively colder and later coires, so will the deer follow and spend more and more of their time on the ground they normally inhabit in summer, where, given the grazing, they can find more peace from disturbance by humans, midges, clegs and other winged pests and from the heat of the hotter summer days.

By early June single hinds will begin to be seen dotted here and there, away from the main herds, indication that their

The hind I saw licking at, and rubbing her head on, a creosoted pole.

In May the coats of the stags are shabby and ugly with loose hair coming out in tufts. New antler growth is well advanced in the mature stags.

'time' is near, and in due course the first new calf will be seen. This is a welcome sight in its ever-recurrent charm and in its implication that the year of the red deer is now really beginning anew. Winter is past and ahead lie months of plenty when the Highlands and the red deer, inseparable to those who know and appreciate their uniqueness in a materialistic world, are at their best.

As June wears on, more and more dappled sprites will be seen with the herds and one wonders anew how such seemingly delicate creatures can thrive and be so obviously full of the joys of life amid surroundings which, to the human eye, can often look so forbidding. By the time the red deer calf is only three days old it can outrun any human unwise enough to test his powers against it, and this over the roughest of hill ground. Those apparently fragile thin legs, disproportionately

As the new grazing begins to show higher and higher on the hill so the deer work higher out, following it.

When at long last fresh foliage appears in late May deer will rear up to browse on leaves within reach.

long-looking at that age, are endowed with the resilience, strength and nimbleness more often attributed to wild goats, but equally true of Highland red deer.

For their first two weeks of life the dappled calves will be dependent entirely on milk from their mothers, milk which must be nutritious indeed, for a calf can double its weight in that first fortnight. From then on they will begin, experimentally, then enjoyably, imitating their mothers and other members of the group they will now be with, to sample the novel tastes, each one a fresh delight, of the June grazing, enjoying the sunshine or seeking the shade as the fancy takes them, coming low into the sheltered coires to graze each evening, or when occasional bouts of bad weather dictate, and journeying back up to the high ridges, replete with their overnight grazing, to spend yet another carefree day while weather and grazing are good and disturbance is rare.

One final thought for those who enjoy, or would like to enjoy, watching red deer on the high hills. The ideal is to achieve this *unseen and unsuspected,* so that one can lie as I did only days before penning this, in warm summer sun, on a high ridge overlooking a small group of deer some way below, also lying, quietly chewing the cud. In the group of fourteen there were four dappled youngsters, two lying sound asleep, curled up like dogs, nose to tail, secure in the presence of their elders, while the other two were dozing, dark-polled and long-eared heads now and then drooping drowsily then jerking up in temporary wakefulness again. Refreshed myself, I left them, undisturbed because unconscious of my departure, which was as carefully 'engineered' as my approach, a peaceful scene embodying the tranquillity of the hills themselves. Long may it continue to be so.

The family tie is strong in deer especially on the female side. A hind (foreground) and her calf of the previous year, now almost a year old, still with her although she is heavy in calf again. Both have typically untidy-looking coats of bleached, loosening old winter hair.

Index